WITHDRAWN FROM
TSC LIBRARY

OCEAN THERMAL
ENERGY CONVERSION

UNESCO ENERGY ENGINEERING SERIES
ENERGY ENGINEERING LEARNING PACKAGE

Dr Boris Berkovski
Chairperson of the Editorial Board
Director, Division of Engineering and Technology
UNESCO

Organized by UNESCO, this innovative distance learning package has been established to train engineers to meet the challenges of today and tomorrow in this exciting field of energy engineering. It has been developed by an international team of distinguished academics coordinated by Dr Boris Berkovski. This modular course is aimed at those with a particular interest in renewable energy, and will appeal to advanced undergraduate and postgraduate students, as well as practising power engineers in industry.

Solar Electricity
Edited by Tomas Markvart

Magnetohydrodynamic Electrical Power Generation
Hugo Messerle

Geothermal Energy
Edited by Mary H. Dickson and Mario Fanelli

Energy Planning and Policy
Maxime Kleinpeter

Ocean Thermal Energy Conversion
Patrick Takahashi and Andrew Trenka

Industrial Energy Conservation
Charles M. Gottschalk

OCEAN THERMAL ENERGY CONVERSION

Patrick Takahashi
Hawaii Natural Energy Institute
University of Hawaii
USA

and

Andrew Trenka
Pacific International Center for High Technology Research
Hawaii
USA

JOHN WILEY & SONS
Chichester · New York · Brisbane · Toronto · Singapore

Copyright © 1996 UNESCO

Published in 1996 by John Wiley & Sons Ltd.
Baffins Lane, Chichester,
West Sussex PO19 IUD, England
National Chichester (01243) 779777
International (+44) 1243 779777

All rights reserved.

No part of this book may be reproduced by any means,
or transmitted, or translated into a machine language
without the written permission of the publisher.

*Any opinions expressed in this text reflect exclusively those of the authors
and are not necessarily those of UNESCO or any other affiliated organization*

*UNESCO gratefully acknowledges the financial contributions and other
forms of co-operation received from the International Technological University ITU).
London and Paris, over the period 1991–1994 towards several of the texts in the Series.
Without the consent and goodwill of the ITU Board of Trustees, and especially Dr S. M. A
Shahrestani, one of its founding members, some of the texts could not have been produced.*

Other Wiley Editorial Offices

John Wiley & Sons, Inc., 605 Third Avenue,
New York, NY 10158–0012, USA

Jacaranda Wiley Ltd, 33 Park Road, Milton,
Queensland 4064, Australia

John Wiley & Sons (Canada) Ltd, 22 Worcester Road,
Rexdale, Ontario M9W ILI, Canada

John Wiley & Sons (SEA) Pte Ltd. 37 Jalan Pemimpin #05-04.
Black B. Union Industrial Building, Singapore 2057

Library of Congress Cataloging-in-Publication Data

Takahashi, Patrick K.
    Ocean thermal energy conversion / P. Takahashi, A. Trenka
      p.   cm — (UNESCO energy engineering series)
    Includes bibliographical references and index.
    ISBN 0 471 96009 8
    1. Ocean thermal power plants.   I. Trenka, A.   II. Title.
III. Series.
TK1073. T35 1996
621.44—dc20   95-23823
                                                                CIP

British Library Cataloguing in Publication Data

A catalogue record for this book is available from the British Library

ISBN 0 471 96009 8

Typeset in 10/12pt Times by Pure Tech India Ltd., Pondicherry
Printed and bound in Great Britain by Redwood Books, Trowbridge, Wilts

Contents

Foreword vii

Preface ix

Aims and objectives xi

1 Overview 1
   Aim 1
   Objectives 1
   1.1 Introduction
   1.2 Summary of OTEC research 2
   1.3 Multiple-product OTEC systems 6
      1.3.1 Electrical power generation 7
      1.3.2 Fresh water production 7
      1.3.3 Air-conditioning and refrigeration 8
      1.3.4 Aquaculture and mariculture 8
      1.3.5 Coldwater agriculture 8
   1.4 Integrated OTEC systems 9
   1.5 Future challenges 11
   Self-assessment questions 12
      Answers 13

2 Ocean Energy Technology 17
   Aim 17
   Objectives 17
   2.1 Tidal energy systems 17
   2.2 Wave energy systems 19
   2.3 Current energy systems 22
   2.4 Salinity gradient systems 23
   Self-assessment questions 26
      Answers 28

3 Economics and Externalities 33
   Aims 33
   Objectives 33

3.1 OTEC systems   33
3.2 Tidal energy systems   42
3.3 Wave energy systems   44
3.4 Current energy systems   45
3.5 Salinity gradient systems   46
Self-assessment questions   47
Answers   47

4 OTEC Thermodynamics   49

Aims   49
Objectives   49
4.1 The first and second laws of thermodynamics   50
4.2 Thermodynamics of OTEC power generation   54
4.3 Performance of OTEC components   57
4.4 Fluid dynamics of the OTEC pipeline   60
4.5 OTEC systems   62
    4.5.1 Closed-cycle OTEC   63
    4.5.2 Open-cycle OTEC   69
Self-assessment questions   74
Answers   75

References   79

Index   81

Foreword

Education applied to a whole complex of interlocking problems is the master key that can open the way to sustainable development.

A major obstacle to the development of renewable energies is the paucity of relevant information available to engineers and technicians (who too often lack the necessary know-how and skills) as well as to decision-makers and users. One of UNESCO's priorities in this domain is therefore to promote training and information aimed at sensitizing specialists and the general public to the possible uses of renewable energy sources, with particular regard to environmental concerns and the requirements of sustainable development.

The present textbook on renewable energies in UNESCO's Energy Engineering Series is designed not only to provide instruction for new generations of engineers but also to foster the kind of environmental management so urgently needed throughout the world. The series as a whole is a co-operative venture involving many institutions working together—on the basis of pre-defined standards—to promote awareness of the environmental, cultural, economic and social dimensions of renewable energy issues, as well as knowledge of those technical aspects that form the core of the UNESCO/ Learning Package Programme.

The first High-level Export Meeting, held at UNESCO Headquarters in July 1993 as part of the World Solar Summit Process (WSSP), made education and training one of its strategic priorities. The series should make a useful contribution to the WSSP and UNESCO is pleased to be a partner in its development.

Federico Mayor
Director-General
UNESCO

Preface

This book begins with an overview of ocean energy and integrated ocean resource applications. Ocean thermal energy conversion (OTEC) is treated in greater depth because it is a topic of advanced development in many regions of the world and offers enormous potential for reducing our dependence on fossil fuels within the next decade through the production of electrical power, plus the utilization of the coldwater effluent for integrated applications. Therefore, the focus of this book is on this promising technology, although other ocean energy systems are also discussed at length.

The basic objective is to present the scientific principles and development potential of ocean energy resources. Other objectives are to:

- provide the fundamentals so that extension and integration can be developed for future applications;
- survey the state of the technology to give a sense of timing;
- lay out economic alternatives to ensure realistic projections; and
- suggest the environmental attractiveness of ocean energy as a viable option for future consideration.

The growing concern over global climate change and other environmental problems has recently pushed numerous renewable energy developments close to realization. The particular advantages of OTEC systems include the production of clean, environmentally benign energy, as well as valuable co-products and the possible mitigation of global warming (greenhouse effect) through the absorption of atmospheric carbon dioxide.

This book is the product of the combined effort of several researchers and engineers of the Hawaii Natural Energy Institute (HNEI) at the University of Hawaii and the Pacific International Center for High Technology Research (PICHTR) located in Honolulu. The first three chapters are intended for those interested in a non-technical description of the current developmental status of ocean energy systems. Chapter 4 provides theoretical and technical information concerning OTEC systems for engineers and others interested in obtaining a thorough understanding of the scientific principles involved.

Developing countries should find information from the first three chapters useful, in addition to certain parts of Chapter 4. Hands-on training and computer application packages can be provided in Hawaii should a field course be desired.

Aims and objectives

Each chapter opens with a statement of its aims and objectives. Those of Chapter 1 are to present an overview of ocean thermal energy conversion (OTEC), consisting of an explanation of the concept of OTEC with a brief discussion of the different types of OTEC systems, i.e., closed-cycle, open-cycle, and hybrid-cycle, a review of the history of OTEC research to date, a summary of the multiple products that can be obtained from OTEC, and some of the challenges involved in the large-scale development and widespread use of this form of renewable energy.

To understand the potential, limitations, and operation of OTEC requires familiarity with some elementary concepts of engineering thermodynamics and fluid mechanics. Chapter 2 provides a brief review of these concepts. The objective of this review is to present material pertinent to performance studies of OTEC systems. In consideration of this specialized focus, the breadth of discussions of fundamental principles is restricted, and detailed analyses of the operation of individual system components are omitted.

A summary of the first and second laws of thermodynamics is presented first. These laws then are applied to assess performance of OTEC power generation cycles from the perspective of component and system efficiencies. Finally, since transport of large quantities of seawater through pipelines is a distinguishing feature of OTEC, a relationship to estimate the energy requirements for this process is introduced.

The objectives of Chapter 3 are to present a more detailed discussion of the two predominant OTEC heat engine cycles—the Rankine or closed-cycle, and the Claude or open-cycle—and to provide basic information on four other forms of ocean energy which are more restricted in availability or less developed technologically, i.e., tidal power, wave power, current energy, and salinity gradient energy.

Finally, the objectives of Chapter 4 are to present information on ocean energy economics, such as market potential, capital cost estimates, and production rates for OTEC and other ocean energy systems where available, and to discuss the environmental impacts associated with these systems.

1
Overview

AIM

The aim of this chapter is to understand the limitations and potential environmental benefits of ocean energy compared to fossil fuel powerplants, and to describe the advantages of ocean thermal energy conversion (OTEC), particularly in terms of multi-products and opportunities for open ocean mariculture and marine biomass plantations.

OBJECTIVES

When you have completed this chapter you should be able to:

1. Discuss the ocean energy options available for development.
2. Explain why ocean thermal energy conversion has an economic advantage over other ocean options which only produce electricity.
3. Diagramatically articulate the difference among closed, open, and hybrid OTEC cycles.
4. Speculate on the industrial opportunities that major open ocean ventures could provide in the next century.

1.1 INTRODUCTION

The oceans occupy almost three-quarters of the earth's surface and represent an enormous source of nonpolluting, inexhaustible energy. They can provide an alternative energy source that can be utilized to offset reliance on combustion of fossil fuels and their resultant environmental problems of global warming and air pollution.

While many of the major developed nations have conducted exploratory research and development, and even installed a few commercial facilities, the total operational power available, with the exception of a French tidal power plant, is far less than 100 megawatts. Conversely, the projected available ocean power far exceeds the ultimate energy consumption of mankind, making this option extremely attractive, especially when the environmental implications are considered.

Most of the technology development involving tidal, wave, current, and salinity gradient energy systems has focused intensively on electricity production. Relatively little progress has been made with currents and salinity gradients. Tidal and wave energy systems appear to have good potential for further refinement and are within an acceptable range for economic competitiveness. Small-scale OTEC systems providing electrical power, nutrients for mariculture, and fresh water are ideal for expanding the economic potential of many island and coastal communities. This will enable OTEC technology to mature and scale-up to larger systems in the future.

Solar energy is absorbed and stored as heat in the surface layer of the ocean. Cold water is found at depths of 800 meters. The OTEC process uses this temperature difference to convert thermal energy to mechanical energy for the generation of electricity. When the warm surface water temperature and cold deep water temperature differ by at least 20°C, an OTEC system can produce net power. Tropical regions worldwide have a temperature difference of 20°C or greater throughout the year, thus providing an enormous potential for meeting the future energy needs of the world (Figure 1.1).

Research and development efforts have identified several applications for the cold, nutrient-rich, pathogen-free deep seawater in addition to the generation of electricity. The by-products of energy production can expand the economic output of OTEC systems through applications such as mariculture and agriculture, fresh water production, air-conditioning, and refrigeration. The prospects for commercialization of OTEC are promising, in the near term for island communities and in the longer term for open ocean grazing platforms supporting integrated production systems (Takahashi et al., 1992).

The promise of environmental 'enhancement' is also possible through the induced upwelling of deep ocean water. The potential to sequester atmospheric carbon dioxide through the combined effect of biological growth and minerals addition is rapidly becoming an important consideration. In the meantime, open ocean fisheries and marine biomass plantations can produce revenues to support the total system economics (Takahashi and Trenka, 1992).

1.2 SUMMARY OF OTEC RESEARCH

The concept of ocean thermal energy conversion was first proposed by the French engineer Jacques Arsene d'Arsonval more than a century ago. He

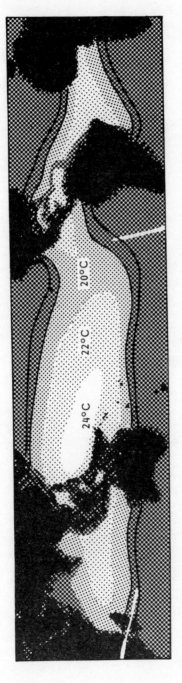

Figure 1.1 Temperature difference between the surface of the ocean and deep water in tropical regions

envisioned a closed-cycle heat engine using ammonia as the working fluid, but he never tested the concept. Almost fifty years passed before one of d'Arsonval's students, Georges Claude, designed and field-tested an experimental system at Matanzas Bay in northern Cuba in 1930. Although his model generated 22 kilowatts (kW) of power, it consumed more power than it generated. Claude later attempted a floating plant aboard a cargo vessel moored off the coast of Brazil. Waves destroyed the coldwater pipe as it was being deployed, and Claude never achieved his goal of generating net power with an OTEC system (Penny and Bharathan, 1987).

A French team designed a 3 megawatt (MW) plant for the west coast of Africa in 1956, but the plant was never built. A private consulting engineer, J. Hilbert Anderson, and his son began a serious design analysis of OTEC in the 1960s. William E. Heronemus of the University of Massachusetts and Clarence Zener of Carnegie-Mellon University joined them in the early 1970s. The National Science Foundation awarded a grant to the University of Massachusetts in 1972 to assess the technical and economic feasibility of the OTEC process. They awarded another grant the following year to Carnegie-Mellon to investigate other elements of the OTEC system (Committee on Alternate Energy Sources, 1975).

When the oil embargo of 1973 prompted an international search for alternative sources of energy, the potential of OTEC was reexamined. In 1979 a closed-cycle Mini-OTEC plant produced net power, and closed-cycle heat exchangers were tested by a vessel designated as OTEC-1 in 1980. Both projects were conducted in Hawaii. Additional experiments in Hawaii at the Seacoast Test Facility of the Natural Energy Laboratory of Hawaii (NELH) on the western shoreline of the Big Island of Hawaii showed that cold deep ocean water is rich in nutrients and relatively pathogen-free. Commercial development resulted from this work, and NELH has added the Hawaii Ocean Science and Technology (HOST) Park for new enterprises to continue developing applications for cold seawater.

There are three basic types of OTEC cycles now under development. The closed-cycle system (Figure 1.2(a)) uses the warm surface seawater to evaporate a working fluid such as ammonia or Freon, which drives a turbine generator. After passing through the turbine, the vapor is condensed in a heat exchanger cooled by water drawn from the deep ocean. The working fluid is pumped back through the warm water heat exchanger, and the cycle is repeated continuously (National Oceanic and Atmospheric Administration, 1984).

The open-cycle system (Figure 1.2(b)) uses warm surface seawater as the working fluid. The warm water is pumped into a flash evaporator in which the pressure has been lowered by a vacuum pump to the point where the warm seawater boils at ambient temperature. The steam produced drives a low-pressure turbine to generate electricity. The steam is then condensed in a heat exchanger cooled by deep ocean water, producing desalinated water as a by-product (Solar Energy Research Institute, 1989).

SUMMARY OF OTEC RESEARCH 5

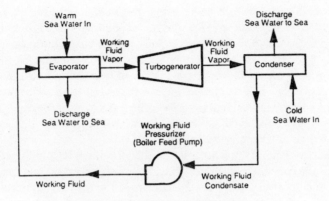

(a) Schematic of a **closed-cycle** OTEC system

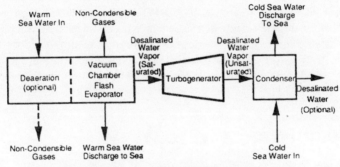

(b) Schematic of an **open-cycle** OTEC system

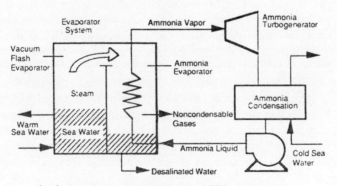

(c) Schematic of a **hybrid-cycle** OTEC system

Figure 1.2 OTEC cycles

The hybrid-cycle has not been tested, but in theory, this type of system (Figure 1.2(c)) combines the principles of the open- and closed-cycle systems, maximizing the use of the thermal resource by producing both electricity and desalinated water. First, electricity is generated in a closed-cycle stage. The temperature difference in the seawater effluent from the closed-cycle stage is sufficient to produce desalinated water by using a flash evaporator and a surface condenser in a second stage. Another possibility is the use of a second stage with an open-cycle system, which should double the output of desalinated water (Penny and Bharathan, 1987).

In 1986 the Pacific International Center for High Technology Research (PICHTR) began the open-cycle OTEC research program. Test results from research conducted by the US Department of Energy on evaporator spouts, warm seawater deaeration, mist removal, condensation capacity, and other factors using apparatus for measuring heat and mass transfer at NELH supported the design and construction of an open-cycle net power-producing experiment (NPPE) in Hawaii. Ground-breaking for the NPPE took place in November 1991. In the spring of 1993, NPPE generated 200 kW gross power and is now producing operational and experimental data necessary to scale up to a larger commercial-size plant. The program is supported by the US Department of Energy and the State of Hawaii (Rogers and Trenka, 1989). A private consortium in 1993 began constructing a similar, but closed-cycle, pilot plant at NELH.

1.3 MULTIPLE-PRODUCT OTEC SYSTEMS

Early OTEC experiments focused on energy systems, but the cold deep ocean water, which is relatively pathogen-free and rich in nitrates, phosphates, and silicates, was soon identified as a resource with other potential. For the past three years, PICHTR, with the support of the Government of Japan through the Ministry of Foreign Affairs, has been designing a multiple-product OTEC system (Figure 1.3). A multiple-product OTEC (MP-OTEC) system not only generates electricity, but also provides desalinated water, air-conditioning and refrigeration, and permits mariculture and agriculture operations in an integrated system. This development is important for island nations that need energy, water, and food to improve their economies and quality of life. Work at the NELHA/HOST Park has stimulated commercial development through the use of cold ocean water for mariculture, refrigeration, and air-conditioning. There are now five coldwater pipes for experimental work at NELHA in addition to lines constructed by private companies.

A survey in 1987, initiated by PICHTR, investigated locations in the Pacific region as sites for deployment of an MP-OTEC system. Of the thirty islands and Asian locations included in the survey, PICHTR's analysis of populations, economies, internal policies, energy demands and projections, and OTEC-related cash-crop potential, found eight locations with high potential:

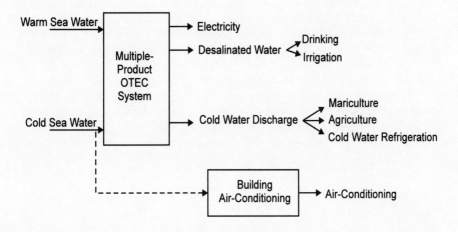

Figure 1.3 A multiple-product open-cycle OTEC system

American Samoa, Western Samoa, the Cook Islands, Tonga, Guam, Belau (Palau), Pohnpei, and the Commonwealth of the Northern Mariana Islands. A separate PICHTR project also assessed Christmas Island in the Republic of Kiribati as another location for future OTEC development. PICHTR completed a conceptual design of a 1 MW open-cycle OTEC plant for Pacific island application in 1989. A net power output of 1 MW corresponds to the electricity demand of many small Pacific island communities. A primary criterion for shore-based plants is that a source of deep coldwater must be within a few kilometers of the plant site, since the coldwater pipe is a major factor in determining the cost of an OTEC plant.

1.3.1 Electrical power generation

One-megawatt OTEC plants could presently be cost-competitive in remote, oil-dependent Pacific island countries. Fuel costs currently amount to more than one-half the total income from imports in these areas. The levelized cost of electricity from a 1 MW OTEC plant over a 30-year period is estimated at $0.11 to $0.19/kWh. Since the cost of electricity ranges from $0.16 to $0.44/kWh in isolated island communities, significant savings could be realized through the use of OTEC, in addition to the other benefits offered by this technology.

1.3.2 Fresh water production

The desalinated water produced by open-cycle and hybrid-cycle OTEC systems is actually purer (less saline) than the water provided by most municipal water systems. Estimates indicate a 1 MW plant fitted with a second stage fresh water production unit could supply approximately 55 kilograms per second of fresh water, approximately 4750 m^3/day, sufficient for serving a population of 20 000. Fresh water production from reverse osmosis and multi-stage flash desalination plants costs between $1.30 and $2.00/m^3 for a plant with a 4000 m^3/day capacity. Using these figures, a 1 MW OTEC plant could produce almost $3 million worth of desalinated water per year. In addition to potable, fresh water for domestic use, desalinated water from OTEC can be used for crop irrigation to increase food supplies.

1.3.3 Air-conditioning and refrigeration

Compared with conventional air-conditioning equipment, cold seawater used to air-condition buildings can result in substantial savings. The cold seawater can circulate through space heat exchangers or can cool a working fluid circulated through heat exchangers. PICHTR research estimates that a 300-room hotel can be air-conditioned by the cold seawater piped to a 1 MW OTEC plant at less than 25% of the cost of electricity for operating a conventional air-conditioning system. The pay-back period for the capital investment of installing a cold seawater air-conditioning system is estimated to be four years or less. The coldwater can also be used to create cold storage space for preserving seafood and other products.

1.3.4 Aquaculture and mariculture

Many new strategies involving seafood and other products grown in seawater pumped from the ocean have been proposed and tested in Hawaii. OTEC-related mariculture ventures at NELH and the associated HOST Park now represent more than $50 million (US) in capital investment. Several coldwater pipes are in place, the largest of which has a diameter of nearly one meter, producing fluids from a depth of 700 meters. Current ventures include culturing salmon, abalone, American lobster, flat-fish, sea urchin, edible seaweed such as ogo and nori, and algae for chemical extracts.

The Kochi Artificial Upwelling Laboratory near the Murota Peninsula on Shikoku Island, Japan, features a 12.5 centimeter (diameter) coldwater pipe, bringing coldwater from 320 meter depths. Growth kinetics for marine plants and animals are being conducted. Although OTEC mariculture is still in the developmental stage, steady progress is being made. Further research and commercialization activities are needed to demonstrate its economic viability.

1.3.5 Coldwater agriculture

Researchers of the University of Hawaii were first to propose the idea of using cold seawater in agriculture. An array of coldwater pipes buried in the ground creates cool weather growing conditions not found in tropical environments. The system also produces drip irrigation by atmospheric condensation on the pipes. Using this method, the growth of strawberries and other spring crops and flowers throughout the year in the tropics has been demonstrated. Commercial developers have initiated 'coldwater' agriculture enterprises in Hawaii.

1.4 INTEGRATED OTEC SYSTEMS

The long-term value of ocean energy systems will no doubt be tied to multi-purpose integrated applications. These will range from 1 MW and smaller sizes with a range of co-products, to, ultimately, floating cities and industrial complexes powered by gigawatt-sized plants. The earlier applications for artificial upwelling might well use surge pumps instead of OTEC, as there is sufficient nutrient content from depths of several hundred meters.

Through the remainder of this decade, system applications will follow the model operating at Keahole Point in Hawaii: land-based for power, fresh water, air-conditioning, and mariculture products. The next generation systems early in the 21st century will move into the open ocean on floating platforms, where the coldwater effluent will be mixed with the warmer surface water so that the high nutrient fluid can be maintained in the euphotic zone for photosynthesis. Closing the biological growth cycle will result in sea ranches for pelagic species such as tuna and mahimahi, and marine biomass plantations which can provide the feedstock for conversion to transportation fuels such as methanol.

An international partnership of industry, government, and academia to design, build, and operate a large floating platform powered by ocean thermal energy conversion would be an imaginative venture of commercial potential with possible environmental benefits. Such a platform would also:

- provide fresh water, air-conditioning, and artificially upwelled fisheries and marine biomass plantations;
- be serviced by a variety of ocean robotics;
- demonstrate the feasibility of seabed ore and methane clathrate mining and processing at sea;
- advance the prospects of tapping oceanic hydrothermal deposits;
- conduct tests to create new materials and sensors to withstand corrosion and deep ocean conditions; and
- serve as a marine scientific base for experiments and measurements.

Figure 1.4 illustrates the proposed program in a total systems framework.

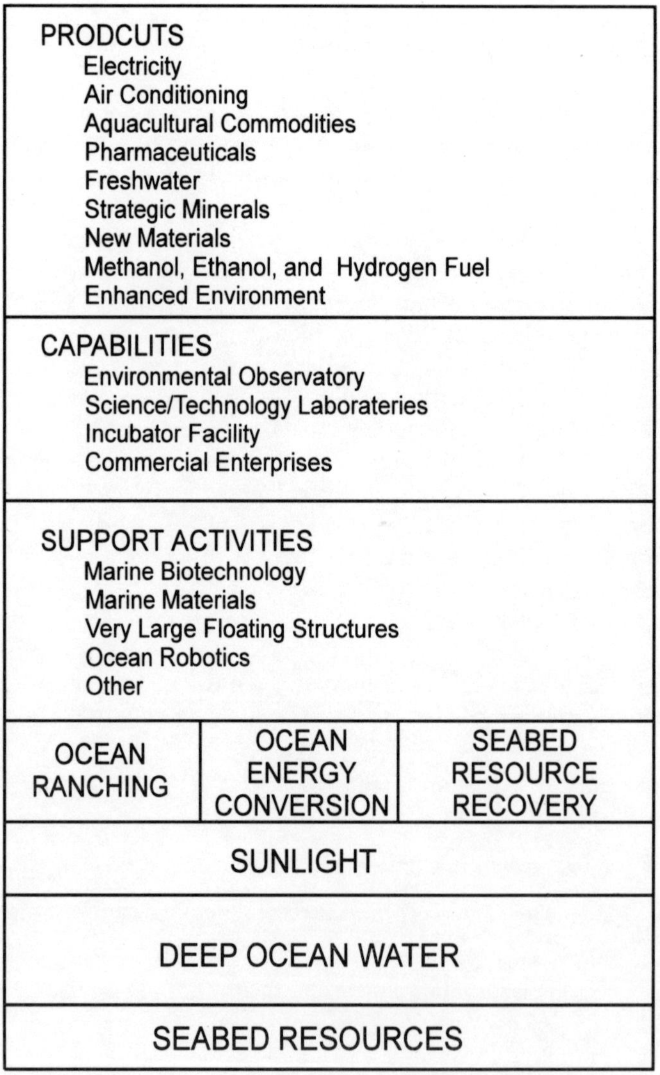

| PRODCUTS |
| Electricity |
| Air Conditioning |
| Aquacultural Commodities |
| Pharmaceuticals |
| Freshwater |
| Strategic Minerals |
| New Materials |
| Methanol, Ethanol, and Hydrogen Fuel |
| Enhanced Environment |

| CAPABILITIES |
| Environmental Observatory |
| Science/Technology Laboraterics |
| Incubator Facility |
| Commercial Enterprises |

| SUPPORT ACTIVITIES |
| Marine Biotechnology |
| Marine Materials |
| Very Large Floating Structures |
| Ocean Robotics |
| Other |

| OCEAN RANCHING | OCEAN ENERGY CONVERSION | SEABED RESOURCE RECOVERY |

| SUNLIGHT |

| DEEP OCEAN WATER |

| SEABED RESOURCES |

Figure 1.4 Integrated ocean resource development facility

The field of ocean resource technology has developed to the point where a major new imaginative program is warranted. The Apollo Project cost more than $20 billion to reach the moon. While the dollars spent did much for national pride, there were few commercial benefits. NASA recently broke ground at Cape Canaveral for the $50 billion space station program, which appears to be in considerable financial difficulty; however, the 'space' consortium has captured the imagination of the general public, something that the ocean community has failed to achieve so far. The National Oceanic and Atmospheric Administration recently hosted a planning session to look at the potential of building up to 500 OTEC grazing plantships both to generate revenues through the commercialization of ocean resources, but also to prevent or retard hurricane formation through careful placement in those portions of the ocean where storms initially develop. Part of the motivation was in response to defense conversion needs related to innovative use of shipyards. Perhaps this grander vision of ocean projects can begin to gain the interest of the general public.

1.5 FUTURE CHALLENGES

Installations of commercial OTEC plants at various sites throughout the world by the year 2010 are predicted. A combination of technical and economic factors create challenges in realizing the widespread commercial utilization of OTEC systems.

To construct future plants as large as several hundred megawatts, as conceived by OTEC planners, much larger-diameter seawater pipes must be available. One proposed solution is the use of large pipes made of a flexible membrane that would be 'inflated' by the positive pressure of seawater supplied by pumps mounted in the coldwater zone on the sea bottom. However, bottom-mounted pumps present a difficult maintenance problem. Also suggested is the mounting of OTEC plants on 'towers' resting on the sea bottom, which would serve as a framework for a very large-diameter pipe.

Much larger turbines than those in current usage will be required for use with the low-pressure steam of the OTEC process. These new turbines will have to be constructed of lightweight, durable materials. PICHTR's research plan calls for testing innovative turbine designs.

PICHTR estimates a very good sales potential for 1 MW (net output) OTEC plants for the Pacific area. However, financing methods for countries with marginal economies need to be worked out. One approach to financing OTEC projects in the Pacific region involves PICHTR's assistance to governments in gathering, verifying, and analyzing data to develop business plans for seeking financing from appropriate agencies.

Mariculture operations based on OTEC technology that can be economically viable must be thoroughly pursued. The potential to produce food products for local consumption and export will essentially lower the cost of operating an OTEC plant. Establishing mariculture operations in some island locations may be very inexpensive because the addition of nutrient-rich seawater to lagoons or bays could stimulate biological activity as well as attract marketable species from the open ocean.

A significant challenge lies in the logistics of deploying OTEC plants in remote locations. Some island communities have inadequate harbors, airports, and housing, and lack technical expertise in operation and maintenance and other elements needed to accommodate OTEC plant installation and operation. Plants could be delivered in a special OTEC deployment vessel containing all the components, equipment, and personnel necessary for complete installation. After the installation, local residents could be trained in operation and maintenance.

During the past 15 years research and development of OTEC systems have produced a wealth of data on the scientific and technical aspects of the technology. Research continues to refine OTEC knowledge. The NPPE and other small experimental plants and pilot-scale commercial plants planned in the next few years could lead to widespread commercialization of OTEC by the year 2010.

Analysis shows that the key to commercial success of OTEC does not lie solely in the generation of electricity, but depends heavily on the development of other potential uses of clean, cold seawater. OTEC by-products such as fresh water, air-conditioning and refrigeration, mariculture and agriculture, and other innovative applications offer great promise for OTEC as a total resource system, as is being demonstrated in Hawaii today.

SELF-ASSESSMENT QUESTIONS

1 Give a brief history of research in ocean thermal energy conversion (OTEC) and describe the status of OTEC at present.

2 Describe the components and physical processes of three types of OTEC systems.

3 What is meant by 'multiple products of OTEC?' Give examples of the major products.

4 List and discuss some of the challenges involved in the large-scale commercial development of OTEC.

5 List the primary forms of ocean energy available and discuss the origin of each.

6 OTEC power is best produced in the extreme northern and southern latitudes because the surface waters are colder. True or false?

7 Ocean energy, per kilowatt hour of electricity generated, produces far less carbon, nitrogen, and sulfur oxides than fossil fuel plants. True or false?

8 The source of power for any ocean energy system can be created

   (a) only by the sun?

   (b) only by the moon?

   (c) by both the sun and/or the moon?

   (d) by neither the sun nor the moon?

9 The economic attractiveness of OTEC is related to

   (a) the low cost of electricity?

   (b) the abundance of co-products to produce a flexible revenue stream?

   (c) the high cost of the powerplant?

   (d) all of the above?

Answers

1 The OTEC concept was first proposed by d'Arsonval in the late 1800s. Georges Claude tested a system in Cuba in 1930, but never achieved net power generation. A small amount of research was performed in the following decades by various researchers, and the potential of OTEC was more seriously reexamined after the oil embargo of 1973. In Hawaii in 1979 a closed-cycle Mini-OTEC plant produced net power, and closed-cycle heat exchangers were tested in 1980. This was followed by the development of mariculture operations in Hawaii using cold deep ocean water rich in nutrients.

A Japanese consortium built and tested a closed-cycle power plant on Nauru in 1981. From 1982 to the present, Japanese organizations have conducted OTEC simulations and experimental plant projects. Research conducted by the US Department of Energy supported the design and construction of an open-cycle net power-producing experiment in Hawaii to produce operational and experimental data necessary to scale up to a larger commercial-size plant. Thus, OTEC is in the pre-commercial stage, although several companies have proposed building commercial plants in various parts of the world.

2 (a) *Closed-cycle OTEC* The closed-cycle system uses the warm surface seawater to evaporate a working fluid such as ammonia or Freon, which drives a turbine

generator. After passing through the turbine, the vapor is condensed in a heat exchanger cooled by water drawn from the deep ocean. The working fluid is pumped back through the warm water heat exchanger, and the cycle is repeated continuously.

(b) *Open-cycle OTEC*   The open-cycle system uses warm surface seawater as the working fluid. The warm water is pumped into a flash evaporator in which the pressure has been lowered by a vacuum pump to the point where the warm seawater boils at ambient temperature. The steam produced drives a low-pressure turbine to generate electricity. The steam is then condensed in a heat exchanger cooled by cold deep ocean water, producing desalinated water as a by-product.

(c) *Hybrid-cycle OTEC*   The hybrid-cycle combines the principles of the open- and closed-cycle systems. First, electricity is generated in a closed-cycle stage. The temperature difference in the seawater effluent from the closed-cycle stage is sufficient to produce desalinated water by using a flash evaporator and a surface condenser in a second stage. Another possibility is the use of a second stage with an open-cycle system, which should double the output of desalinated water.

3 A multiple-product OTEC system not only generates electricity, but also provides desalinated water, air-conditioning, and refrigeration, and permits mariculture and agriculture operations in an integrated system.

4 To construct OTEC plants with a capacity of more than a few megawatts, larger-diameter seawater pipes must be developed, and much larger turbines will be required. Financing methods for countries with marginal economies need to be created. Successful mariculture based on OTEC technology is needed. Logistics of deploying OTEC plants in remote locations will have to be worked out.

5 Ocean thermal energy conversion (OTEC): temperature differential between surface and deep ocean water.

- Tidal: variation of tides
- Waves: waves and surges
- Currents: movement of ocean currents
- Salinity gradient: salinity concentration differential between fresh and salt water

6 False. OTEC is best produced at the equator (at $\pm 20°$ latitude) where the temperature differential between surface and deep water is the greatest. The combination of geo-fluids (warm) and arctic/antartic surface waters can also produce OTEC power; however, this combination is not the *best*.

7 True. The upwelled fluids can result in the release of CO_2 into the atmosphere, but

this amount is small compared to that from fossil plants and can be reduced through bio-production.

8 (c) The moon and sun create tidal power, while the sun is responsible for the others.

9 (b) Co-products make OTEC economically promising.

2
Ocean energy technology

AIM

The aim of this chapter is to describe the fundamentals of various ocean energy systems and to present the current developmental status of these technologies.

OBJECTIVES

When you have completed this chapter you should be able to:

1. Discuss why tidal powerplants fluctuate in peak power production.
2. Explain the advantages and disadvantages of single and double basin tidal energy systems.
3. Given the deep water wave height and period, determine the wavelength, wave frequency, and power per unit of wave crest width generated.
4. Calculate maximum current power that can be extracted from an ideal, non-ducted turbine.
5. Explain how salinity gradients can be used to produce power.

2.1 TIDAL ENERGY SYSTEMS

Tidal energy is derived from the enormous energy induced in the oceans by the gravitational forces of the sun, moon, and earth. The ebb and flow of the powerful ocean tides are greatly influenced by the tidal swing, the volume of water involved, and the inshore geological features. Tidal energy is generated by collecting rising tidal water behind a barrier and then releasing it at ebb tide through turbines to generate electricity. Systems are available to extract energy by relatively conventional hydroelectric turbines and related structures.

The use of conventional technology separates the development of tidal energy from other ocean energy sources. Low-head, axial-flow turbines are the modern means of harnessing the relatively small differences in water level in a river system or from a reservoir. Tidal energy extraction requires a strong ocean effect and a natural resonant inshore configuration to make it work most efficiently and economically. The energy is as predictable as the state of the ocean tide at any instant.

Tidal energy extraction appears to be less demanding on advanced technologies than other energy sources; however, it highly depends on natural processes. Just as hydroelectric power depends on natural differences in the terrain elevation, tidal power depends on the natural configuration of inshore geological features. Few coastal areas exist where conditions combine to produce the degree of resonance required. Natural sites and celestial forces appear to favor the development of tidal power systems within latitudes of 50 to 60 degrees (Warnock, 1987).

There are very few tidal energy power stations operating in the world today. The four countries with functioning systems are France, the former Soviet Union, China, and Canada. These countries have been the most actively involved in the study of tidal energy conversion. The total power generated by these systems is about 263 MW.

The La Rance tidal power station on the north-west coast of France, with an installed capacity of 240 MW, is the world's largest. Construction began in 1961 and was completed by 1968. The powerhouse is 390 meters long, housing twenty-four 10 MW turbine units. The plant has been operational since 1968 with an outstanding 95% availability.

In 1968, the Soviet Union put into operation a 400 kW pilot plant in Kislayan Gulf, called the Kislogubskaya pilot plant; it pioneered floating construction techniques. In 1985, the Soviet Union announced plans to build a tidal plant on the White Sea coast with a generating capacity of 15 000 MW. Called the Mezenskaya plant, it is still in the preliminary stages of development.

In the People's Republic of China, tidal energy was first reported in 1959 with the installation of a 40 kW plant located in Shashan. A 165 kW tidal plant was later built in 1970 in the Shandong Province on the Jingang Creek. In May 1980, China's first two-way tidal plant, rated at 500 kW, began operating on the Jiangxia Creek near the Zhousan Islands. This plant was later expanded to 3.2 MW in 1986.

The Canadian tidal power project is the 20 MW plant at Annapolis Royal, Nova Scotia. This plant was built by the Nova Scotia Power Corporation. Located on the Annapolis River near its outlet at the Bay of Fundy, the plant contains a single 'Straflo' hydropower turbine and has been operational since 1984. Presently, Canada is investigating the potential of such turbines for

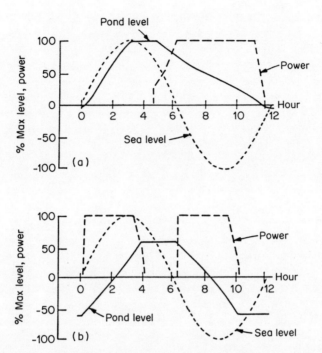

Figure 2.1 Variation of sea level, basin level, and power output for single basin schemes: (a) single-effect; (b) double-effect (EPRI, 1986). Copyright © 1986, Electric Power Research Institute, EPRI AP-4921. *Ocean Energy Technologies: The State of the Art*. Reprinted with permission

larger-scale installations in the Bay of Fundy, and for low-head run-of-the-river developments.

A typical relationship among sea level, basin water level, and power production is illustrated in Figure 2.1. Note that generation is possible only when the available head (difference between basin and sea level) exceeds a certain threshold. This results in generation during less than 50% of the time. At the beginning and end of the power phase, when head is low, generation is less than capacity, while during higher head, constant output is achieved by regulating turbine flow. Because the moon's position relative to the earth varies over a period of 24 hours and 50 minutes, tidal phase shifts by almost an hour each day. Hence generation occurs at different times on different days and may not always coincide with peak electrical demand.

2.2 WAVE ENERGY SYSTEMS

Ocean wave energy conversion technologies utilize the kinetic energy of ocean waves to produce power. Wave energy is a potential environmentally benign and renewable energy resource. The general approaches to converting wave energy into electricity can be broadly categorized by means of deployment and means of energy extraction and conversion. Means of deployment include floating deep-water technologies and shallow-water, fixed-bottom technologies. Means of energy extraction and conversion include mechanical cams, gears, and levers; hydraulic pumps; pneumatic turbines; oscillating water columns; and funnelling devices (Hay, 1990). Currently, five wave energy systems generate a total of 535 kW of power, and two more commercial systems are expected to be operating in the near future. Various wave energy devices are depicted in Figure 2.2.

The Japanese government has had a very active wave energy research and development program for many years. Applications under investigation range from wave power generators for lighthouses and light buoys to wave pump systems, ship propulsion, and energy for road heating, heat recovery systems, and fish farming. Several technologies have been examined by both government and industry under the Japanese wave energy program. These include floating terminator-type wave devices, fixed coastal-type wave power extractors, and applications of oscillating water column turbines. The best known project, supported by the International Energy Agency, was the Kaimei, a 500 ton barge containing about ten oscillating water columns. Wave action produces oscillations of the water column that produce pneumatic power which is converted to electrical power via air turbo-generators.

The Indian government has investigated the potential of oscillating water column plants and studied the feasibility of building a 5 MW plant in a new harbor breakwater near Madras. Presently, a 150 kW demonstration plant using a Wells turbine-generator is under construction at a fishing harbor near the port of Trivandrum.

The United Kingdom wave energy program was initiated in 1974. Ongoing development projects include wave-powered desalination and pumping, investigation of the use of a Wells turbine in naturally formed rock gullies, construction of a 75 kW prototype wave power plant on the Scottish island of Islay, production of wave-powered turbine generators for navigational buoys in Northern Ireland, and development and model testing of a small-scale Sea Clam wave energy converter at Loch Ness by Coventry University.

Norway has conducted an extensive wave power program since 1975. In the 1980s, these efforts included the installation by Kvaerner Brug of a 500 kW prototype wave power system, called the multiresonant oscillating water column (MOWC), on the west coast of Norway. Operational since November 1985, the plant was swept off its foundation and destroyed during a severe storm in January 1989.

In 1986, the Norwegian firm Norwave installed a new system called Tapchan, a tapered channel wave powerplant, in Bergen, Norway, that produces 350 kW of power. Typically, a tapered channel is carved out of a rocky coastal area, using shaped charges if necessary. The taper can handle a wide spectrum of wavelengths efficiently. As a wave passes through the tapered channel, its wave height is gradually increased as the channel narrows. The

WAVE ENERGY SYSTEMS 21

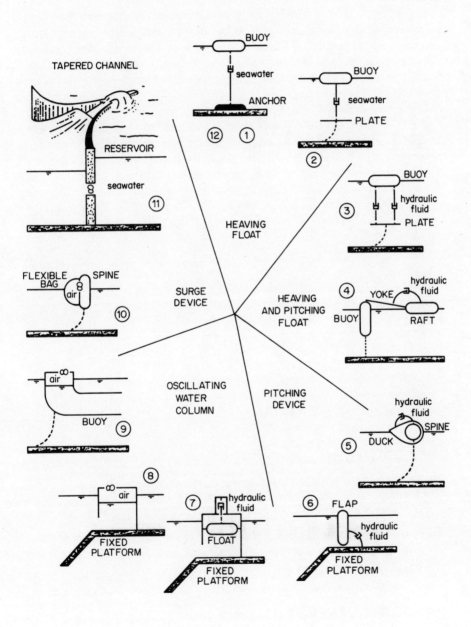

Figure 2.2 Stylized representation of wave energy devices (Hagerman and Heller, 1988) Reproduced by permission of SEASUN Power Systems.

wave then spills over into a reservoir where it is stored and subsequently passes through a low-head Kaplan water turbine to generate electricity.

In Sweden, use of a heaving buoy as a wave energy converter has been extensively studied. Field tests of a 30 kW prototype hose-pump device have recently been completed off Sweden's west coast and there was a proposal for a 1 MW pilot plant to be installed off the Atlantic coast of Spain. Pharos Marine in the United Kingdom has developed a wave-powered navigation buoy using the same concept.

In the former Soviet Union, testing was done for a 3 kW wave powerplant and a 50 kW inertial wave power unit at Makhachkala in the Caspian Sea. In the United States, wave energy activity has included research and development of the McCormack pneumatic turbine; prototype testing of a heaving and pitching circular float, tandem-flap system, and contouring raft device; and research on a heaving buoy type, wave-powered desalination system in Puerto Rico (Saris et al., 1989).

The relationship between the wave period, T, and the wavelength, L, is:

$$L = \frac{gT^2}{2\Pi} = 1.56T^2 \text{ (meters)}$$

where g is the gravitational acceleration $= 9.81$ m/s^2.

Small amplitude waves in deep water will, according to linear wave theory, have a power per unit of wave crest width of:

$$P = \frac{\rho g^2 H^2 T}{32\Pi} \approx 0.98 H^2 T \text{ (kW/meter)}$$

where

ρ = water density (kg/meter^2)
g = gravitational acceleration (m/s^2)
H = wave height (meters)
T = wave period (seconds)

The corresponding relation for random seas (irregular waves) is:

$$P = 0.55 H_s^2 T_z \text{ (kW/meter)}$$

where H_s = significant wave height (the average wave height of the largest third of the observed waves), and T_z = average time interval between successive crossings of the mean high water level.

2.3 CURRENT ENERGY SYSTEMS

The kinetic energy of river currents has been used from medieval times to produce power using simple water turbines. There are many old prints that show mechanical power produced from mills at bridges to pump river water to the adjacent communities. The proposed application of current turbines in the oceans is a comparatively recent development, and has been prompted by the observations of mariners and oceanographers of the swiftly flowing current in some regions of the world. The Gulf Stream, or more specifically the Florida Current, is of particular interest because of the high current velocity and its proximity to large centers of population on the Florida coast. The Florida Current is particularly strong off the city of Miami, and ocean current turbines have been proposed to utilize this resource.

The performance of an ocean current turbine is similar to the performance of a wind turbine. The ocean or wind turbine transforms a proportion of the kinetic energy of the flow into mechanical power. A small ocean turbine was demonstrated in 1985 in the Florida Current. The unit was suspended from a research vessel at a depth of 50 meters and developed approximately 2 kW. The project was privately funded, and a proposal was made to design and test 100 kW and 1 to 2 MW units of a similar design (EPRI, 1986).

In addition, a 20 kW prototype turbine, designed by UEK Corporation, is under research and development, for which testing is planned in New York City's tidal East River. Since 1979, Canadian researchers at Nova Energy Ltd have been developing large Darrieus-type vertical axis turbines for hydropower applications and are presently completing testing of a 5 kW prototype. Australian current energy conversion units designed by Tyson Turbines Ltd are small to medium-size modular devices capable of producing an energy output of more than 670 kW depending on depth and stream velocity. These units are commercially available for a variety of applications and have been demonstrated in many countries including Australia, the Philippines, Mexico, the United States, and Canada (Saris et al., 1989).

From simple momentum analyses, it can be shown that the maximum power that can be extracted from an ideal, non-ducted turbine occurs when the turbine reduces the stream velocity to one third of the freestream value, hence

$$\text{Power(kW)} = C_p(\tfrac{1}{2}\rho V^3 S)$$

where

C_p = power coefficient (maximum = 0.59)
V = fluid velocity (m/s)
S = turbine disk area (m^2)
ρ = fluid density (kg/m^3)

A configuration for a proposed open current turbine is shown in Figure 2.3.

2.4 SALINITY GRADIENT SYSTEMS

A large unused source of energy exists at the interface between fresh water and salt water, and the extent of energy depends on the salinity gradient. In

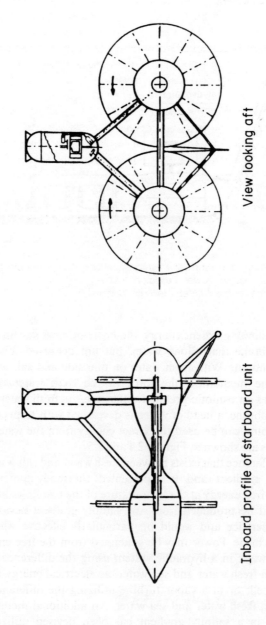

Figure 2.3 Arrangement of a proposed open current turbine (EPRI, 1986) Copyright © 1986, Electric Power Research Institute, EPRI AP-4921. *Ocean Energy Technologies: The State of the Art*. Reprinted with permission

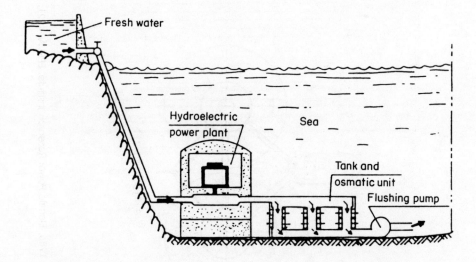

Figure 2.4 Arrangement of a submarine hydro-electric-osmotic power plant (EPRI, 1986). Copyright © 1986, Electric Power Research Institute, EPRI AP-4921. *Ocean Energy Technologies: The State of the Art.* Reprinted with permission

extracting this salinity gradient energy, the heart of most systems is a semi-permeable membrane that allows water, but not dissolved solids, to pass through the membrane. With fresh water on one side and salt water on the other side of the membrane, the force of the fresh water through the membrane creates an osmotic pressure difference. As fresh water permeates through the membrane, a head of water is developed with respect to the salt water, and a turbine can be used to extract energy from the water flow. Two proposed systems are shown in Figures 2.4 and 2.5.

The energy difference that exists between fresh water and salt water depends on the salinity gradient and is represented thermodynamically as the difference in the free energy at the temperature of the two flows of water. The power that could be produced from any salinity gradient device increases with salinity difference and would be particularly effective when the salt water is a dense brine. Power may be generated from the free energy difference in various ways: in a hydraulic system using the difference in osmotic pressure between fresh water and sea water; as electrical energy in a reverse electro-dialysis cell; or in a vapor turbine utilizing the difference in vapor pressure between fresh water and sea water. An additional method of using the free energy in a salinity gradient has been devised utilizing the extension and contraction of special fibers induced by changes in salinity (EPRI, 1986).

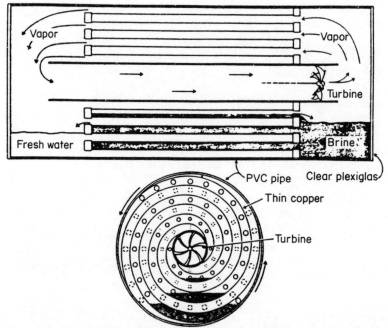

Figure 2.5 Arrangement of a laboratory model of a vapor pressure salinity gradient power unit (EPRI, 1986) Copyright © 1986, Electric Power Research Institute, EPRI AP-4921. *Ocean Energy Technologies: The State of the Art*. Reprinted with permission

SELF-ASSESSMENT QUESTIONS

1. List four types of ocean energy technologies other than OTEC and describe their status of technological development and commercial usage.
2. From the basic theory of tidal power, for a simple single-pool tidal system, the average potential power for one tidal period is (Twidell and Weir, 1986):

$$P_{avg} = \frac{\rho A R^2 g}{2T}$$

where

   R = tidal range
   A = area of basin
   T = tidal period

   Calculate the average potential power for a simple single-pool tidal powerplant that has a basin area of 13 000 km² and a tidal range of 8 meters. Assume a tidal period of 6 hours and 12.5 minutes, and a 27.5% efficiency.

3. Water is pumped rapidly from the ocean at high tide to give an increased water level in a tidal power basin of 1 meter. If the tidal range is 5 meters and if the pump/

generator system is only 50% efficient, show that the extra energy gained can be nearly twice the energy needed for pumping (Twidell and Weir, 1986).

4 Describe the advantages of tidal energy extraction with respect to other ocean energy sources.

5 What are some of the difficulties associated with tidal energy extraction?

6 The single basin scheme and the double basin scheme are two principal ways in which power can be generated from a tidal estuary. What are some of the advantages and disadvantages of using two basins for tidal energy extraction?

7 A deep water wave has a height of 2 meters and a period of 6 seconds. Determine the wavelength, wave frequency, and the power per unit of crest width generated by the wave.

8 The power in a deep water wave having a wavelength of 100 meters is approximately 73 kW/m. Determine the amplitude and period of the wave.

9 Discuss, with respect to period and amplitude, the type of deep ocean wave that is desired for power generation.

10 List some of the advantages of wave energy systems.

11 Determine the maximum power that can be extracted from an ideal non-ducted turbine placed in a tidal current of 2 meters/second. The diameter of the turbine disk is 2 meters and the power coefficient is 0.59. Assume the water density is 1025 kg/m^3.

12 A non-ducted turbine placed in a tidal current of 3 meters per second develops 20 kW of power. Assuming a power coefficient of 0.40, calculate the required diameter of the turbine disk.

13 List the advantages of utilizing a vertical axis machine to extract energy from ocean current streams.

14 Discuss the advantages of suspending an ocean current turbine midway between the surface and the seabed.

15 Discuss the difficulties involved in using a submerged turbine to extract energy from ocean current streams.

16 Given that the osmotic pressure difference between sea water and fresh water is approximately 22 atmospheres (Taylor, 1983), what is the theoretical power that could be generated if river water flows into the ocean at a rate of 1 m^3/second?

17 Illustrate how the salinity gradient can be used to generate power via a hydroturbine.

18 Describe how the temperature gradient in a fresh water lake differs from the temperature gradient in a salt water lake.

Answers

1 *Tidal energy*: Operating in only four countries at present. Requires special geologic features and tidal range. Power generation may not coincide with peak demand due to lunar cycle.
Wave energy: Still in R&D stage. Many different devices proposed for extracting energy from waves. Few commercial systems operating.
Current energy: Still in the conceptual stage. Some prototypes tested. Commercial potential limited to areas with strong currents.
Salinity gradient energy: Feasibility unproven. Semi-permeable membrane development needed. Very limited commercial potential at present.

2 $$P_{avg} = \frac{(1025 \text{ kg/m}^3)(13\,000 \text{ km}^2)(8 \text{ m})^2(9.81 \text{ m/s}^2)}{(2)(22\,350 \text{ s})}$$
$$= 187\,158 \text{ kg-km}^2/\text{s}^3 = 1.87 \times 10^{11} \text{ watts}$$

Since the efficiency is 27.5%:

$$P_{avg} = (1.87 \times 10^{11} \text{ W})(0.275) = 5.147 \times 10^{10} \text{ watts} = 51.5 \text{ gigawatts}$$

3 Consider a mass M of pumped water. Taking the height of water (h) at the center of mass, the input energy to pump, 1.0 meter is

$$\frac{Mgh}{\text{Eff}} = \frac{Mg(0.5 \text{ m})}{0.50} = Mg$$

The output energy at low tide is

$$\text{Eff} \times Mgh = 0.50 \times [Mg(5.5 \text{ m})] = 2.75\, Mg$$

Therefore

$$\frac{\text{Energy gain}}{\text{Energy input}} = \frac{2.75\, Mg - Mg}{Mg}$$
$$= \frac{2.75 - 1}{1} = 1.75$$

4 The advantages of tidal energy extraction are:

(a) predictability of the energy source;

(b) less demand on advanced technologies than other energy sources.

5 Some of the difficulties associated with tidal energy extraction are:

 (a) Tidal energy extraction requires a strong ocean effect and a natural resonant inshore configuration. Few coastal areas exist where conditions combine to produce the degree of resonance required.

 (b) Turbines must operate at low head with large flow rates–an uncommon condition in conventional hydropower practice.

 (c) Generation occurs at different times on different days and will not, in general, coincide with peak demand.

6 The major advantage of the double basin scheme is that power generation can be timed to take place during the hours of peak demand by using the basins alternately. However, the added complexity results in higher costs and lower output overall.

7 Wavelength: $L = 1.56T^2 = 1.56(6 \text{ s})^2 = 56.2 \text{ meters}$

 Frequency: $f = 1/T = 1/6 \text{ s} = 0.17/\text{second}$

 Power: $P = 0.98H^2T = 0.98 (2 \text{ m})^2 (6 \text{ s}) = 23.5 \text{ kW/meter}$

8 Period: $T = (L/1.56)^{1/2} = 8 \text{ seconds}$
 $H = (P/0.98T)^{1/2} = (73/0.98 \times 8)^{1/2} = 3 \text{ meters}$

 Amplitude: $a = H/2 = 3/2 = 1.5 \text{ meters}$

9 From the relationship

$$P = 0.98H^2T$$

it is apparent that the power in the wave is proportional to the square of the height and to the period of motion. Therefore, long period and large amplitude waves are the best for power generation.

10 The advantages of wave power are:

 (a) it is environmentally benign;

 (b) large energy fluxes are available;

 (c) the ability to forecast wave conditions over periods of days.

11 Maximum power, $P = C_p(0.5\rho V^3 S)$

$P = 0.59[0.5 \times 1025 \text{ kg/m}^3 \times (2 \text{ m/s})^3 \times \Pi/4 \ (2 \text{ m})^2] = 7.6 \text{ MW}$

12 Turbine disk diameter (D):

$P = 0.40[0.5 \times 1025 \text{ kg/m}^3 \times (3 \text{ m/s})^3 \times \Pi/4 \ (D)^2] = 20\,000 \text{ watts}$

$$D^2 = (50\,000 \text{ N-m/s})/[0.5 \times 1025 \text{ kg/m}^3 \times (3 \text{ m/s})^3 \times \Pi/4]$$

$$D = (4.601 \text{ m}^2)^{1/2} = 2.145 \text{ meters}$$

13 The advantages of using a vertical axis machine to extract energy from ocean current streams are:

   (a) high efficiency compared with drag dependent devices;

   (b) vertical axis machines have a low solidity, which reduces cost.

14 The advantages of suspending an ocean current turbine midway between the surface and the seabed are:

   (a) reduced wave-induced loadings;

   (b) loss of output and shear effects resulting from reduced flow near the seabed.

15 The difficulties involved in using a submerged turbine to extract energy from ocean current streams are:

   (a) hazard to shipping if submerged devices are used on large scale;

   (b) difficulties of inspection and maintenance;

   (c) fouling and subsequent power output reduction;

   (d) materials and corrosion problems.

16 The free energy lost when a volume (V) of pure water is mixed with a larger volume $= PV$, where P is the osmotic pressure of the solution. Therefore,

$$\text{Power} = [(22 \text{ atm})(14.7 \text{ psi/atm})(6894 \text{ Pa/psi})](1 \text{ m}^3/\text{s}) = 2.23 \text{ MW}$$

17 As shown in the figure below, the fresh water in pool A is separated from the salt water in pool B by a semi-permeable membrane. The fresh water passes through the membrane, enters pool B, and raises its level. The resulting head of water can then be used to drive a conventional water turbine to generate electric power.

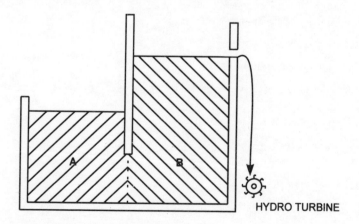

HYDRO TURBINE

18 When sunlight shines on a fresh water lake, the solar energy is received and absorbed below the surface. The water heated by the sunlight becomes less dense and rises to the surface. This results in a convective circulation that maintains cooler water at the bottom and transports the solar energy back to the top. At the surface, the solar energy is dissipated to the atmosphere by conduction, convection, and evaporation. When sunlight shines on a salt water lake, the temperature gradient is reversed. Salt water lakes contain a non-uniform vertical distribution of salt concentration with higher concentrations at the bottom than at the top. This causes the water at the bottom to have higher density and to remain at the bottom, even though it may have a higher temperature. The solar energy absorbed in the deep layers is trapped there because the effect of salt on density offsets the effect of thermal expansion.

3
Economics and externalities

AIMS

The aims of this chapter are to determine the economics of ocean energy, provide insights on present and future markets, understand the environmental implications of ocean energy production, and introduce the concept of externalities and the importance they can play in decision-making.

OBJECTIVES

When you have completed this chapter you should be able to:

1. Describe the two types of markets for OTEC in size, temporal, and financial terms.
2. Calculate the cost of electricity produced by one or more ocean energy systems.
3. Discuss the positive and negative aspects of ocean energy on the environment.
4. Explain how externalities can improve the economics of ocean energy.
5. Determine what breakthrough can make salinity gradient energy costs comparable to fossil plants.

3.1 OTEC SYSTEMS

There are at least two distinct markets for OTEC: (1) industrialized nations and islands; and (2) smaller or less industrialized islands. Islands with modest needs for power and fresh water could use small open-cycle plants. These plants could be sized to produce from 1 to 10 MW of electricity, and at least 450 000 to 8 million gallons of fresh water per day (1700 to 30 000 m³/day). Hybrid-cycle plants can be used in either market for producing electricity and water. For example, a 40 MW hybrid-cycle plant could also produce as much as 16 million gallons of water per day (60 600 m³/day).

Scenarios under which OTEC might be competitive with conventional technologies, in the production of electricity and water, have been assessed (Vega and Trenka, 1989). The capital cost for OTEC plants, expressed in 1989 $/kW, was established assuming modest engineering development. The relative capital cost of producing electricity ($/kWh) with OTEC, offset by the desalinated water production, was then equated to the fuel cost of electricity produced with conventional techniques to determine the scenarios (i.e., fuel cost and cost of fresh water production) under which OTEC could be competitive. No attempt was made at speculating about the future cost of fuel. It was simply stated that if a situation is represented by one of the scenarios, OTEC would be competitive. Four development scenarios are envisioned in Table 3.1.

One scenario corresponds to small island nations, where the cost of diesel-generated electricity and fresh water is such that a small (1 MW) land-based open-cycle OTEC plant, with water production, would be cost-effective today.

Table 3.1 OTEC market penetration scenarios

Nominal net power (NWe)	Type	Scenario requirements	Scenario availability
1	Land-based OC-OTEC with 2nd-stage additional water production[a]	•$45/barrel of diesel •$1.6/m³ water	South Pacific island nations by 1995
2	Land-based as above	•$25/barrel of fuel oil •$0.85/m³ water *or* •$30/barrel with •$0.8/m³ water	American island territories and other Pacific islands by 2000
40	Land-based hybrid (ammonia power cycle with flash evaporator downstream)[b]	•$44/barrel of fuel oil •$0.4/m³ water *or* •$22/barrel with •$0.8/m³ water	Hawaii, if fuel or water cost doubles by 2000
40	•Closed-cycle land-based •Closed-cycle plant-ship	•$36/barrel •$23/barrel	by 2000

[a] OC-OTEC limited by turbine technology to 2.5 MW modules or 10 MW plant (with four modules).
[b] OC-OTEC or hybrid (water production downstream of closed-cycle with flash evaporator).

A second scenario corresponds to conditions that are plausible in the near future, in, for example, territories like Guam, American Samoa, and the Northern Mariana Islands, where land-based open-cycle OTEC plants rated at 10 MW could be cost-effective if credit is given for the fresh water produced. A third scenario corresponds to land-based hybrid OTEC plants for the industrialized nations market producing electricity through an ammonia cycle and fresh water through a flash (vacuum) evaporator. This scenario would be cost-effective with a doubling of the cost of oil fuel or doubling of water costs, and for plants rated at 40 MW or larger. The fourth scenario is for floating OTEC electrical plants, rated at 40 MW or larger, and housing a factory or transmitting electricity to shore via a submarine power cable. These plants could be deployed throughout the tropical regions of the ocean basins and could encompass a large market.

Since environmental protection has been recognized as a global issue, another important point to consider is the preservation of the environment in the area of the selected site, in as much as preservation of the environment anywhere is bound to have positive effects elsewhere. OTEC definitely offers one of the most benign power production technologies, since no hazardous substances need to be handled, and no noxious by-products are generated; OTEC merely requires the pumping and discharge of various seawater masses, which, according to preliminary studies, can be accomplished with virtually no adverse impact. This argument should be very attractive for pristine island ecosystems, as well as for already polluted and overburdened environments.

One major difficulty with OTEC is not of a technological order: OTEC is capital-intensive, and the very first plants, mainly because of their small size, will require a substantial capital investment. Given the prevailing low cost of crude oil, and of fossil fuels in general, the development of OTEC technologies is likely to be promoted by government agencies rather than by private industry. The motivation of governments in subsidizing OTEC may vary greatly, from foreign aid to domestic concerns.

For the former case, ideal recipient countries are likely to be independent developing nations. If these countries' economic standing is too low, however, the installation of an OTEC plant rather than direct aid in the form of money and goods may be perceived as inadequate help. In addition, political instability could jeopardize the goodwill of helping nations to invest. For the latter case, potential sites belong to, or fall within the jurisdiction of, developed countries: examples include Hawaii, Taiwan, Tahiti, American Samoa, the Northern Marianas, and Guam.

A study performed in 1981 for the US Department of State identified 98 nations and territories with access to the OTEC thermal resource (20°C temperature difference between surface water and deep ocean water) within their 200 nautical mile exclusive economic zone (EEZ). A representative list is reproduced in Table 3.2. For the majority of these locations, the OTEC resource is applicable only to floating plants (arbitrarily assuming that the

36 ECONOMICS AND EXTERNALITIES

Table 3.2 Nations and territories with access to the OTEC thermal resource

Geographical Area	Mainland	Island	
Americas	Mexico Brazil Colombia Costa Rica Guatemala Honduras Panama Venezuela Guyana Suriname French Guiana (FR) Nicaragua El Salvador Belize United States	Cuba Haiti Dominican Rep. Jamaica Virgin Is. (US) Grenada St.Vincent Grand Cayman (UK) Antigua (UK) Puerto Rico (US) Trinidad & Tobago Bahamas	Guadeloupe (FR) Martinique (Fr) Barbados Dominica St. Lucia St. Kitts (UK) Barbuda (UK) Montserrat (UK) The Grenadines (UK) Curacao (NETH) Aruba (NETH)
Africa	Nigeria Ghana Ivory Coast Kenya Tanzania Congo Guinea Sierra Leone Liberia Gabon Benin Zaire Angola Cameroon Mozambique Eq. Guinea Togo Somalia	SaoTome & Principe Ascension (UK) Comoros Aldabra (UK) Madagascar	

Indian/Pacific Ocean	India	Australia	Indonesia	Trust Territories (US)
	Burma	Japan	Philippines	Northern Marianas
	China	Thailand	Sri Lanka	Guam (US)
	Vietnam	Hong Kong (UK)	Papua New Guinea	Kiribati
	Bangladesh	Brunei	Taiwan	French Polynesia (FR)
	Malaysia		Fiji	New Caledonia (FR)
			Nauru	Diego Garcia
			Seychelles	Tuvalu
			Maldives	Wake Is. (US)
			New Hebrides (UK/FR)	Solomon Is.
			Samoa	Mauritius
			Tonga	Okinawa (JAPAN)
			Cook Is.	Wallis & Futuna Is. (FR)
			American Samoa (US)	Hawaii

length of the cold water pipe for a land-based plant should not exceed 3000 meters). A significant market potential for OTEC (i.e., 577 000 MW of new baseload electric power facilities) was postulated. Unfortunately, now as then, there is no OTEC plant with an operational record available. This still remains the impediment to OTEC commercialization.

The capital costs required to build OTEC plants have been estimated assuming cost reductions via modest engineering development after the design, construction, and operation of demonstration plants. The cost figures expressed in 1989 dollars are summarized in Tables 3.3 and 3.4 for plants rated at 1, 10, and 40 MW with the indicated desalinated water production.

It is assumed that 1 MW land-based plants could be deployed some time after 1995. Their commercialization must be preceded by the installation of a demonstration plant of 1 MW and 3500 m^3 of desalinated water per day production capacity. These plants would be designed utilizing the state-of-the-art bottom-mounted cold water pipe technology (i.e., 1.6 meter diameter, high-density, polyethylene pipe).

The design of a 10 MW land-based open-cycle plant would be scaled from the 1 MW demonstration plant with a new design for bottom-mounted cold water pipes. The commercialization of 40 MW land-based plants must be preceded by the design and operation of a 5 MWe closed-cycle demonstration plant. These plants also require the development of larger diameter (> 1.6 m) cold water pipes.

To consider the moored or slowly drifting OTEC plantship, a capital cost of 4600 \$/kW (net) is estimated for an electrical production of 380×10^6 kWh (slightly higher than for the land-based plants because of lower pumping power requirements). The cost differential between the moored and the drifting vessels is insignificant at this level of discussion because the savings in mooring system and power cable costs are offset by the propulsion and positioning requirements, as well as by the product transport, for the factory ship. These plants would be designed utilizing the methodology already available for cold water pipes suspended from a vessel.

The following formula, proposed by the Electric Power Research Institute, is used to calculate the production cost of electricity levelized over the assumed life for the OTEC plant (30 years):

$$\text{Production cost (\$/kWh)} = \frac{(FC)(CC) + (OM)(G)(CR)}{(NP)(CF)(8760)}$$

where

FC = annual interest charge (10%)
CC = plant capital cost (\$)
OM = operation and maintenance (annual expenditures)
G = present worth factor (20 years)
CR = capital recovery factor (0.09)
NP = net power production (kW)

Table 3.3 Levelized (amortized) cost of electricity with desalinated water credit at 0.4 $/m³

Plant CF = 80%	Production electricity (kWh)	Production water (m³/day)	Capital cost ($/kW-net)	Fixed charge	OM (% of capital)	Levelized electricity cost with water production credit ($/kWh)
1 MW OC-OTEC Land-based	8.10E+06	1 700	18 200	10%	1.7%	0.27
1 MW OC-OTEC Land-based 2nd Stage	7.30E+06	3 560	23 000	10%	1.7%	0.31
10 MW OC-OTEC Land-based	7.00E+07	15 000	10 700	10%	1.5%	0.14
10 MW OC-OTEC Land-based 2nd Stage	6.30E+07	30 000	14 700	10%	1.5%	0.17
40 MW CC-OTEC Land-based	3.36E+08	0	6 000	10%	1.5%	0.10
40 MW CC-OTEC Land-based 2nd Stage	2.80E+08	60 600	9 400	10%	1.5%	0.12
40 MW CC-OTEC Floater	3.80E+08	0	4 600	10%	1.5%	0.08

Table 3.4 Levelized (amortized) cost of electricity with desalinated water credit at 0.8 $/m³

Plant CF = 80%	Production electricity (kWh)	Production water (m³/day)	Capital cost ($/kW-net)	Fixed charge	OM (% of capital)	Levelized electricity cost with water production credit ($/kWh)
1 MW OC-OTEC Land-based	8.10E+06	1 700	18 200	10%	1.7%	0.24
1 MW OC-OTEC Land-based 2nd Stage	7.30E+06	3 560	23 000	10%	1.7%	0.24
10 MW OC-OTEC Land-based	7.00E+07	15 000	10 700	10%	1.5%	0.11
10 MW OC-OTEC Land-based 2nd Stage	6.30E+07	30 000	14 700	10%	1.5%	0.10
40 MW CC-OTEC Land-based	3.36E+08	0	6 000	10%	1.5%	0.10
40 MW CC-OTEC Land-based 2nd Stage	2.80E+08	60 600	9 400	10%	1.5%	0.09
40 MW CC-OTEC Floater	3.80E+08	0	4 600	10%	1.5%	0.08

CF = production capacity factor (0.80)
8760 = number of hours in one year

The first term simply represents the payment for a fixed interest loan valued at CC over a prescribed term expressed in hourly payments, where the loan is for a plant rated at a power of NP. The second term models the levelized cost of operating and maintaining the plant over the term. Normally, the OM costs are not levelized.

For open- or hybrid-cycle plants, fresh water credit is obtained by multiplying the unit price by the yearly production and substracting the result from the numerator of the expression given above. For the sake of completeness, costs estimated in this fashion are given in Tables 3.3 and 3.4 for unit prices of water at 0.4 $/m^3 and 0.8 $/m^3, respectively, with the OM expressed as a percentage of capital and unlevelized. These estimates illustrate the importance of the water revenue for the small plants (1 to 10 MW), especially with the unit price of water at twice the present rate.

In addition to the fresh water produced by the open-cycle, OTEC offers potential for mariculture co-products and can provide the chiller fluid for air-conditioning systems. The cold seawater contains large quantities of the nutrients required to sustain marine life. Organisms already grown in this environment in Hawaii include algae, seaweeds, shellfish, and fin fish. Although a number of species have been identified as technically feasible, further work is required to identify cost-effective culture methods for the available markets. OTEC mariculture is in its formative years and not ready for commercialization, or to be transferred to nations under development.

The relatively small portion of the cold seawater can also be used as the chiller fluid for air-conditioning systems. A system based on this concept is presently utilized at NELH for one of the buildings. However, this application is seen to be of some significance for small OC-OTEC plants, but is insignificant for larger plants. With the exception of the relatively small use of the cold seawater as AC chiller fluid, OTEC should be considered for its potential production of electricity and desalinated water.

Like any offshore or shoreline project, commercial OTEC facilities will affect the marine environment. Construction activities may temporarily disrupt the sea bed, destroying habitats and decreasing subsurface visibility. Platforms and marine subsystems may attract fish and other marine species, and maintenance routines to reduce biofouling may increase the level of toxic substances. Intake pipes can draw marine organisms through the plant and move large amounts of nutrient-rich water up from deep depths. However, OTEC systems can be designed and located to minimize their potential effects on the environment and even enhance the surrounding ocean, as shown by Mini-OTEC, when augmented fish catches were reported near the site off the Kona Coast of the Big Island of Hawaii.

The construction of land-based or shelf-mounted OTEC plants can disturb the sea bed. Deployment of moorings, cables, pipes, piles, and anchors may churn up the bottom and increase the number of particles suspended in the water. This type of disturbance can affect areas of special ecological importance, such as coral reefs, seagrass beds, spawning grounds, and commercial fisheries. Short-term disruption of most of these habitats is often reversible, as shown by experience with offshore oil rig construction. OTEC developers can minimize disruption by locating plants away from critical habitats. Where necessary, cables and pipes can be routed through natural breaks in near-shore reefs.

OTEC plants discharge large quantities of ocean water and could potentially affect natural thermal and salinity gradients and levels of dissolved gases, nutrients, trace metals, carbonates, and turbidity. If cold- and warm-water streams are mixed and discharged at the surface, the density of OTEC plant discharges will be different from that of the surrounding water. Behavior of the discharge plume will respond to initial discharge momentum and to buoyancy forces that result from initial density differences. Within several hundred meters of the discharge point, the plume will be diluted by ambient ocean water, sink (or rise) to reach an equilibrium level, and lose velocity until the difference between its velocity and ambient current velocity is small. OTEC plants can be designed to stabilize the discharge plume below the mixed layer to protect the thermal resource and to minimize potential environmental effects. However, a key point to consider is that the intelligent management of this discharge could stimulate living marine resources.

An environmental impact of a more serious nature could occur if ammonia, Freon, or some other environmentally hazardous working fluid was accidentally spilled from a closed-cycle OTEC plant. The effect of ammonia on marine ecosystems would depend on the rate of release and the nature of nearby sea life. Small quantities of ammonia probably would stimulate plant growth downstream. A large ammonia spill would be toxic; for example, a 40 MW plant could release enough ammonia to destroy marine organisms over an area as large as four square kilometers (SERI, 1989).

3.2 TIDAL ENERGY SYSTEMS

The United Kingdom conducted a design study of a tidal plant at Langstone Harbour, Hampshire, that had a 3 meter mean annual tidal range generating 24.3 MW on the ebb tide cycle. Construction cost was estimated at $2813/kW and electricity produced at $0.16/kWh in 1986 dollars. In constant 1990 dollars, this equates to about $3511/kW and electricity cost at $0.20/kWh. Figure 3.1 presents some projected energy costs.

Both positive and negative environmental effects can be expected from the development of tidal power plants. While potential effects would be very site-specific, they may be grouped into several categories based on the physical changes brought about by construction and operation of the plant. These

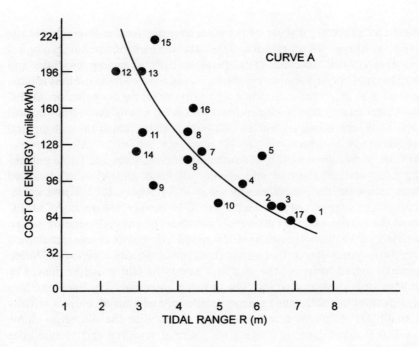

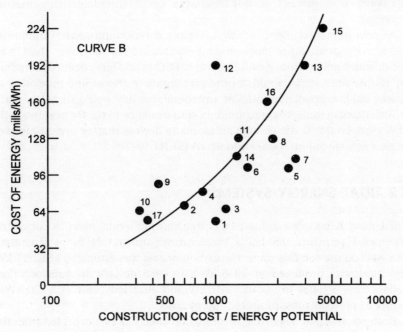

Figure 3.1 Variation of energy costs (1985 dollars) with tidal range and cost factors for 17 sites in the UK (EPRI, 1986) Copyright © 1986, Electric Power Research Institute, EPRI AP-4921. *Ocean Energy Technologies: The State of the Art*. Reprinted with permission

include the physical presence of the dam, changes in water level, changes in flow patterns and current velocities, and changes in sediment patterns.

During construction, dredging operations (including disposal), blasting, and placement of rock fill or concrete structures will impact benthic habitats, increase turbidity (thus affecting organisms within the water column), and may restrict navigation. These impacts would be short-term and local, and would vary depending on whether float-in or cofferdam construction were employed. Once constructed, the physical presence of the dam represents permanent changes which could affect recreational use of the water and impoundment, navigation, and fish passage and habitat. Locks can be used to assist navigation and, in some cases, navigation within the basin could be improved by higher average water levels. The opportunity to build a road across the dam would be a major positive benefit, especially at large sites. The generally smaller tidal ranges, as discussed below, would offer increased opportunity for recreational boating. Because of the large volumes of water involved, entrainment of fish in plant turbines could be a problem. For example, shad migrate through the Bay of Fundy and concern over their entrainment has been cited in regards to potential tidal power projects (EPRI, 1986).

3.3 WAVE ENERGY SYSTEMS

Norwave is developing two commercial Tapchan systems, one to be installed in Java and the other in Tasmania. Each system will produce about 1.5 MW of power. The construction costs range from about $2000/kW in Java to $3550/kW in Tasmania and the systems are expected to produce power at rates of about 5 to 10 cents per kilowatt hour (Vadus, 1991).

The impact on the environment of wave energy conversion is strongly dependent on the scale of the activity. When a modest project is proposed, where the average power delivered to the grid is 40–100 MW, the impacts are likely to be small. However, there will be community resistance should recreational sites be compromised. If a large scheme delivering several thousand megawatts is proposed, the impacts are obviously expected to be larger, and may not be benign.

The conversion of wave energy to electricity may be expected to influence the coastal wave and current climate, the populations of fish and marine mammals, the navigation of ships, and the visual environment. A large wave power conversion system would modify the local wave climate. A reduction in the wave energy arriving at the shores can change the density and balance of species of organisms around the coast, and may modify the deposition of sand on the beaches. The wave energy conversion devices might be expected to have an influence on the populations of fish and marine mammals. Bottom-feeding fish and shellfish, such as lobster and crab, are likely to be unaffected. Fish and marine mammals that spend much of their life near the surface require more consideration. Salmon, members of the herring family, and even sea lions have been mentioned as species that will have to be evaluated when impacts of large-scale wave energy conversion systems are considered.

Wave energy converters placed in or near shipping lanes would present a hazard to shipping because their relatively low profile would make them less visible to sight and radar. The devices would have to be properly marked, and navigation channels would have to be provided through large arrays of the converters. Mooring failures resulting in drifting of the floating devices would provide an additional hazard to navigation and shoreline structures.

The visual effects of wave energy will depend on the site selected, size of the floating platforms, the length of the array, distance offshore, and method of cable transmission. Shore-based systems such as Tapchan may be blended into the coast and require a relatively low-profile reservoir ashore to provide the necessary head of water. Depending on the power, the profile of an array can be barge-like (e.g., the Kaimei Project) and could require a long line of such structures. The visual effect will depend on the distance offshore and the impact is difficult to generalize because each configuration and installation plan will differ from one location to another. Wave energy conversion devices can provide additional benefits such as providing calm seas behind by the breakwater effect, and co-generation of fresh water by forcing seawater through a semi-permeable membrane.

3.4 CURRENT ENERGY SYSTEMS

The Gulf Stream carries 30 million cubic meters of water per second, more than 50 times the total flow in all of the world's fresh water rivers; the surface velocity sometimes exceeds 2.5 meters per second. The extractable power is about 2000 watts per square meter and would, therefore, require extremely large, slow-rotating blade turbines operating like windmills. The total energy of this Florida current is estimated to be about 25 000 megawatts.

In 1979, the Aeroenvironment Company conducted a conceptual design study (Coriolis Project) based on installing very large diameter turbines (referred to as Coriolis turbines) in the Gulf Stream. Energy calculations indicated that an array of 242 large turbines, each about 170 meters in diameter, moored in the Gulf Stream in an array occupying an area of 30 kilometers cross-stream and 60 kilometers downstream would produce about 10 000 MW. This is the energy equivalent of about 130 million barrels of oil per year. Cost estimates indicated that each unit could be built and installed at about $1200/kW in 1978 dollars. Including capital, operating, maintenance, and fuel costs, power is delivered at about $0.04/kWh in 1978 dollars. These figures assumed a plant factor of 57%, which is computed in a way similar to that used for wind turbines, by considering the seasonal variation in the current, plus a two-week annual maintenance shutdown. The Coriolis system is an environmentally benign, cost-efficient method of extracting energy from a renewable source (Lissaman, 1979).

The environmental effect of an array of Coriolis ocean turbines on the Florida Gulf Stream current has been investigated for several models. The results showed that for an annual average extraction of 10 000 megawatts, the reduction in speed of the Gulf Stream is estimated at about 1.2%, much less than its natural fluctuation. Further calculations indicated that any heating effects resulting from turbulence in the wake of the turbines would be very small (Lissaman, 1979). A 1 meter diameter turbine with compliant blades and rim-driven system was constructed and demonstrated in a water flume. No further research and development were conducted. However, more research is needed to provide greater confidence in technical and economic feasibility in constructing, installing, and mooring very large turbines of the size proposed. Current energy systems do not appear to be ready for commercial application at this time.

3.5 SALINITY GRADIENT SYSTEMS

The development of candidate systems for the production of power from salinity gradients has not progressed far enough to provide an accurate economic assessment of system types and configurations. However, general considerations can be presented which point to one concept which may be promising if the salinity gradient is very large, such as where the River Jordan flows into the Dead Sea. An analysis and preliminary experiments for a 100 MW plant at the mouth of the River Jordan indicate that power could be produced at a cost of $0.07/kWh in 1976 dollars (Monney, 1977).

Although this cost is more than double the cost of electricity from a coal-fired power plant, dramatic improvements may be possible for the salinity gradient powerplant with improvements in semi-permeable membranes. A brine such as that which exists at the Dead Sea or Great Salt Lake can be used to produce a greater salinity gradient. Another difficulty is that geographical areas with naturally occurring bodies of high salinity brine are usually deficient in the fresh water needed to provide the salinity gradient. However, it may be possible to use seawater or other brackish water as the low salinity permeate. It may even be feasible to create a renewable energy resource by using seawater in evaporating ponds in a coastal area to produce the high salinity brines which would be mixed with the low salinity seawater permeate.

In an evaluation of the subsystems and components which would comprise a salinity gradient powerplant, it becomes apparent that the semi-permeable membrane is the major technical problem. In all other respects, the plant would draw upon well-established technical capabilities which have little potential for marked improvements. The membrane, however, has a very significant potential for improvement in terms of cost and performance and, at the same time, is the major controlling factor in determining the power output of a plant operating with a specified salinity gradient. The major problems with respect to semi-permeable membrane development are flux, fouling, salt rejection, life expectancy, and cost. The production of power from the salinity gradient between fresh water and seawater will not be economically feasible unless the membrane flux can be improved by an order of magnitude and the requirement for pretreatment of the water can be virtually eliminated.

SELF-ASSESSMENT QUESTIONS

1 In what kinds of locations does the present market exist for OTEC plants?

2 Fresh water can be produced by:
   (a) closed-cycle OTEC plants?
   (b) open-cycle OTEC plants?
   (c) hybrid-cycle OTEC plants?

3 State the formula for calculating the cost of producing electricity by OTEC. Calculate the cost per kWh of electricity from a 1 MW plant costing $20 million with a life of 30 years and annual operation and maintenance cost of $200 000.

4 List the environmental effects expected from the development of tidal energy.

5 List the environmental effects expected from the development of wave energy.

6 In what location does development of current energy systems seem most promising? Why?

7 What developments are required in order for the production of power from the salinity gradient between fresh water and seawater to become economically feasible?

Answers

1 Markets for OTEC plants exist in:
   (a) industrialized nations and large islands;
   (b) smaller and less industrialized island nations.

2 (b) and (c).

3 Production cost of electricity ($/kWh) = $\dfrac{(FC)(CC) + (OM)(G)(CR)}{(NP)(CF)(8760)}$

$= \dfrac{(0.10)(20\ 000\ 000) + (200\ 000)(20)(0.09)}{(1000)(0.80)(8760)} = \$0.34/\text{kWh}$

4 Environmental effects of tidal energy systems:

   (a) During construction, dredging operation, blasting, and placement of rock fill or concrete structures will impact benthic habitats and increase turbidity, and may restrict navigation.
   (b) Presence of the dam represents permanent changes that could affect recreational use of the water and impoundment, navigation, and fish passage and habitat.
   (c) Changes in water level, flow patterns, current velocities, and changes in sediment patterns will occur.

4 Environmental effects of wave energy systems:

   (a) Large wave energy system would modify local wave climate.
   (b) Reduction in wave energy arriving at shores may change the density and balance of species of organisms around the coast and may change the deposition of sand on the beaches.
   (c) Wave energy systems may influence the fish and marine mammal population.
   (d) Wave energy systems near or in shipping lanes may present hazards to shipping.

6 Location of current energy extraction systems in the Gulf Stream seem to hold the most promise. The Gulf Stream carries 30 million cubic meters of water per second, which is more than 50 times the total flow in all the world's fresh water rivers. The total energy of the Gulf Stream current is estimated to be approximately 25 000 megawatts.

7 The semi-permeable membrane is the major technical problem associated with salinity gradient power extraction. The production of power from the salinity gradient between fresh water and seawater will not be economically feasible unless the membrane flux can be improved by an order of magnitude and the requirement for pretreatment of water can be virtually eliminated.

4
OTEC thermodynamics

AIMS

The aims of this chapter are to provide the mathematical and scientific fundamentals necessary to calculate efficiencies and conduct comparative analyses of OTEC systems, to gain familiarity with some elementary concepts of thermodynamics and fluid mechanics in order to determine the potential limitations and operation of OTEC systems and their components, and to understand the engineering principles necessary to design and operate an OTEC system.

OBJECTIVES

When you have completed this chapter you should be able to:

1 Differentiate between the first and second laws of thermodynamics and provide a comparative example of this difference.

2 Determine the Carnot cycle efficiency for one or more OTEC systems.

3 Describe the three classes of turbomachinery proposed for OTEC systems.

4 Discuss what developments needed to be made in cold water pipe and turbine components for 100 MW and larger OTEC plants to be cost-effective.

5 Explain the difference between the Rankine and Claude cycles.

6 Write the expression for the thermal efficiency of a Rankine cycle in terms of properties at various points of the cycle.

4.1 THE FIRST AND SECOND LAWS OF THERMODYNAMICS

Engineering thermodynamics is a field of study concerned with transformations or exchanges of energy within a system or between a system and its surroundings. Here, the term system denotes either a fixed quantity of matter, a device, or a region in space that is being examined. Four laws constitute the basis for all engineering thermodynamics analyses. The so-called zeroth law of thermodynamics states that any two objects in thermal equilibrium with a third object (such equilibrium being manifested by zero net heat transfer) are, in turn, in thermal equilibrium with one another. This law is invoked each time a calibrated measurement device is used to determine temperature. The third law of thermodynamics establishes an absolute datum for a property called entropy. The third law is particularly relevant to the study of chemically reacting mixtures.

For OTEC systems, where chemical reactions generally are assumed to play an insignificant role in the power generation process, the first and second laws of thermodynamics are of primary interest to the engineer. Before presenting a statement of the first law, it is useful to review the concepts of properties, state, equilibrium, energy, and work.

Thermodynamic properties are characteristics of a system that can be quantified, typically by reference to some datum or standard, and which are related to the energy of matter. Properties include pressure, temperature, density, or mass and volume, as well as internal energy and entropy. The state of a system is simply a condition described by a given combination of values of its properties. In this respect, the state can be perceived as a location in an N-dimensional space whose coordinates are the different properties; the state is a vector whose components are the relevant properties.

Extensive discussions are available on the concept of equilibrium. For the purpose of this review, it is sufficient to observe that equilibrium is attained in a system isolated from its environment when there is no tendency for a spontaneous change in its state. By isolated, it is meant that no transfers of energy or mass occur between the system and its surroundings. A corollary to the preceding statement is that, for a system in equilibrium with itself, translations in property space, i.e., changes in state, are indicative of exchanges of energy or mass with its environment.

For most engineers, the concept of energy is rooted in Newtonian mechanics. Energy is perceived as something possessed by matter that manifests itself by a perceptible motion or the performance of some task involving overcoming a resisting force, this action in turn being given the name work and the symbol W. The equivalence of work and energy is employed extensively in discussions of thermodynamics.

It is convenient to perform an artificial division of the energy of matter into microscopic and macroscopic components. The microscopic component is referred to as the internal energy and usually is assigned the symbol U.

Internal energy accounts for motion of the molecules or atoms, for example the random thermal motion of gas molecules or the vibrations of atoms in a crystalline lattice, and, for applications involving chemical reaction, also may account for energy stored in molecular bonds. The microscopic motions that contribute to U have zero mean values, i.e., when averaged over the ensemble of molecules or atoms that comprise the system, the resultant velocity and displacement are zero. Internal energy also must take into consideration the distribution of electrons among the allowed quantum levels and relative to the nuclei.

The macroscopic component is subdivided further into kinetic and potential energy terms. The kinetic energy of a quantity of matter depends on the selected inertial reference frame and is proportional to the square of the velocity of the center of mass evaluated in this reference frame. Non-zero potential energy implies that a macroscopic coordinate of the quantity of matter, e.g., its center of mass, magnetic or electric dipole moment, is displaced from a stable condition dictated by a related applied field, e.g., gravitational, magnetic, or electric. For example, potential energy increases as an object is raised above some 'ground level'.

Energy may be transported or transformed from one form to another. The three modes of energy transport that we will consider are convection, heat transfer, and work. Convection occurs when there is a net transfer of mass in space. As mass moves between locations, it carries along energy. Heat transfer occurs in the presence of a temperature gradient; this thermal non-equilibrium forces a redistribution of internal energy. Heat transfer is given the symbol Q. A relationship employed to evaluate the work performed on a system (i.e., energy transfer from the environment to the system) is

$$W = \int \mathbf{F} \cdot d\mathbf{X}$$

where $\mathbf{F}$ is the vector of force applied by the surroundings on the matter in the system and $\mathbf{X}$ is a displacement occurring at the point of application. Since the scalar product of the vectors $\mathbf{F}$ and $d\mathbf{X}$ is taken, W is a scalar. In many cases, $\mathbf{F}$ and $\mathbf{X}$ may not have the dimensions of force and displacement, respectively, e.g., when evaluating the work associated with the rotation of a shaft due to an applied torque or the polarization of a dielectric by an electrical field. For this reason, $\mathbf{F}$ is often referred to as the generalized force and $\mathbf{X}$ as the generalized displacement.

Thermodynamic analyses typically focus on identifying the modes and amounts of energy transfers between a system and its surroundings and assessing the resulting changes in state. Once the boundaries of the system are clearly identified, an energy balance, or accounting, can be performed that considers the transfers that occur across these boundaries and changes in state taking place within them. This accounting can be summarized as (Reynolds and Perkins, 1977):

Energy production within the system = energy outflows into the system
− energy inflows into the system
+ increase in energy within the system

Given the practical benefits of this production bookkeeping formalism, we now offer a simple statement of the first law of thermodynamics: *The production of energy is always and everywhere equal to zero.* Energy, like mass, is a conserved quantity; it may be transferred and transformed but may not be created or destroyed.

Figure 4.1 depicts a system exchanging energy with its environment by all three modes: convection, heat transfer, and work. The system comprises a region in space bounded by an imaginary, closed control surface (indicated by the dotted line). If E is taken to represent the sum of all microscopic and macroscopic energy of the matter in the system at a given time, then, on a rate basis

$$\text{Increase in energy within the system} = \frac{dE}{dt}$$

Outflows and inflows may be tallied by reference to the (assumed) direction of the arrows drawn in the figure. As shown, outflows include the heat transfer terms Q_1 and Q_3 (expressed as energy/time), the power transmitted as W_2, and convection at locations 1 and 4. The rate of inflow of energy must account for Q_2, W_1, W_3, and convection at locations 2 and 3. Note that m_i is the mass

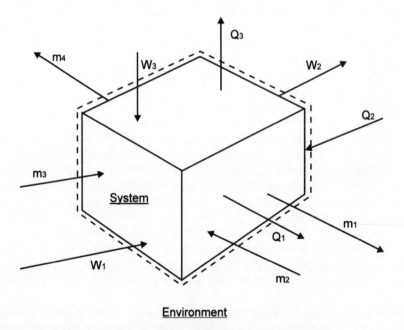

Figure 4.1 System exchanging energy by convection, heat transfer, and work

flow rate (in, say, kg/s), proceeding in the direction of the arrow, at location i. If we assume that equilibrium conditions and one-dimensional flow exist at locations 1–4, and that the influence of electrical and magnetic fields is negligible, then the energy balance becomes

$$0 = [m_1(h_1 + V_1^2/2 + gz_1) + m_4(h_4 + V_4^2/2 + gz_4) + Q_1 + Q_3 + W_2]$$
$$- [m_2(h_2 + V_2^2/2 + gz_2) + m_3(h_3 + V_3^2/2 + gz_3) + Q_2 + W_1 + W_3]$$
$$+ dE/dt$$

where V_i is the magnitude of the velocity of material at i, z_i is the displacement above ground level in the direction of the gravitational field, g is the local gravitational acceleration, and h_i is the specific enthalpy, defined as the sum of the specific internal energy (internal energy per unit mass), u, and the product of the local pressure, P, and specific volume, v (the inverse of density), i.e.,

$$h = u + Pv$$

Enthalpy is a property of matter and is used extensively in analyses of systems in which mass crosses a control surface. The Pv term accounts for the work required to push mass out of the way at these points of crossing. A general expression for the energy balance is

$$\text{Production} = 0 = \sum_{\text{out}} [m_k(h + V^2/2 + gz)_k + Q_k + W_k]$$
$$- \sum_{\text{in}} [m_k(h + V^2/2 + gz)_k + Q_k + W_k] + dE/dt$$

The first summation is taken over all outflows from the system while the second considers all inflows.

Although most of the development behind the preceding example has been omitted, it serves to demonstrate the underlying rationale and general form of the energy balance utilized to analyze the performance of engineering devices. Moreover, it clearly states the constraint imposed by the first law of thermodynamics: that, deprived of a source or sink for energy (i.e., a non-zero production term), the net exchange of energy between a system and its environment is directly related to a change in E and, hence, the state, of the material within the system.

In order for an event or process to occur, an energy balance on the system of interest must satisfy the first law of thermodynamics; however, satisfying the first law is not sufficient to guarantee that a process or event is indeed feasible. Viability must be assessed by invoking the second law of thermodynamics.

In statements of the second law, the property of interest is the entropy, S. A detailed explanation of entropy is too involved to recount here. Entropy should be perceived as a property that quantifies uncertainty about the microscopic condition of matter. Specifically, it is related to the number of ways the total energy of a system may be distributed among the allowed energy states of the ensemble of constituent molecules or atoms. The more energy states accessible, the greater the uncertainty that any specific distribution will exist at a given time. Entropy increases as this uncertainty increases.

Entropy may be transferred in space by convection or heat transfer. Work interactions are not accompanied by a transfer in entropy. One way to explain this is to consider that work involves organized events that are perceived at the macroscopic level. Since entropy reflects randomness at the microscopic level, it can be argued that there can be no relationship between work and entropy. On the other hand, heat transfer occurs as a result of random collisions between molecules and atoms. The end result of these collisions is a transfer of both internal energy and entropy. It can be shown that entropy transfer proceeds at a rate given by

$$\frac{dS}{dt} = \frac{Q}{T}$$

where T is the absolute temperature at the location where heat transfer is taking place.

As in the case of energy, an accounting may be performed on entropy exchanges between a system and its environment. In production bookkeeping form, this balance may be expressed as:

Entropy production within the system = entropy outflows from the system
  − entropy inflows into the system
  + increase in entropy within the system

For a flow-through system similar to that depicted in Figure 4.1, a rate expression for the production of entropy is

$$\text{Rate of production of entropy} = P_s = \sum_{out}\left[ms + \frac{Q}{T}\right]_k - \sum_{in}\left[ms + \frac{Q}{T}\right]_k + \frac{dS}{dt}$$

where s is the specific entropy (per unit mass).

In terms of production bookkeeping, the second law of thermodynamics may be stated as: *The production of entropy must always be greater than or equal to zero* (i.e., $P_s \geq 0$). Unlike energy, entropy may be created and, once created, may never be destroyed. Hence, only processes or events that conserve energy and result in non-negative entropy production are viable. If entropy production is zero, the process or event is called reversible; if greater than zero, irreversible.

4.2 THERMODYMICS OF OTEC POWER GENERATION

The objective with OTEC is to fabricate a system that transforms the internal energy of warm, surface seawater to electrical power which can be delivered to consumers. Since the source of energy stored in warm seawater is the sun, OTEC is a form of indirect solar energy power conversion. A device that converts internal energy to work or power is called a heat engine. Specifically, a heat engine is a system that receives energy from its environment by heat transfer and returns energy to its environment through work. Historically, the study of heat engines has been a primary focus of thermodynamics. One of the major results of classical thermodynamics is embodied in the Kelvin–Planck statement on the operation of heat engines.

A heat engine receives energy from a thermal resource (or reservoir) by means of heat transfer. It can be shown that the second law requires that heat transfer take place from a warmer body to a cooler body. If no energy is rejected from the heat engine to its environment by heat transfer, it is referred to as a 1T (for one thermal reservoir) heat engine. If heat is rejected to a cooler thermal sink, then it is called a 2T (for two thermal reservoirs) heat engine. Figure 4.2 provides a sketch of 1T and 2T heat engines.

Typical heat engines operate cyclically. That is, the state of the material in the heat engine is the same at the beginning and the end of the power production process. This means that, over one complete cycle, the storage of energy within the system is zero. An energy balance for the heat engine, performed over one cycle, demands that either:

$$0 = \Delta W - \Delta Q_h + 0 \text{ (1T heat engine)}$$

or

$$0 = (\Delta W + \Delta Q_l) - \Delta Q_h + 0 \text{ (2T heat engine)}$$

Note that the zero energy production and energy storage terms have been included for completeness. DW and DQ are the total energy transfers, which

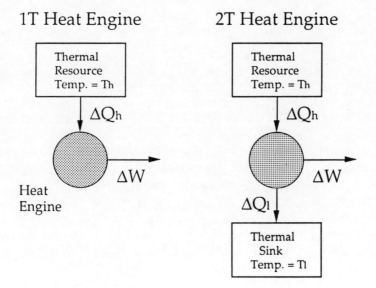

Figure 4.2 1T and 2T heat engines

take place over the course of a cycle, between the engine and its surroundings, by work and heat, respectively. Rearranging terms we have expressions for the work produced:

$$\Delta W = \Delta Q_h \text{ (1T heat engine)}$$

or

$$\Delta W = \Delta Q_h - \Delta Q_l \text{ (2T heat engine)}$$

For a 1T heat engine operating cyclically, all energy received from the thermal resource is converted to work. For a 2T heat engine, only a portion of this energy is converted to work, the balance being rejected to the thermal sink.

The second law of thermodynamics as expressed by the Kelvin–Planck statement imposes a major constraint on the operation of a cyclic heat engine. Briefly, it states that a cyclic 1T heat engine is impossible. Based on the first law balances performed above, this means that 100% conversion of the energy extracted from a thermal resource, DQ_h, to usable work is impossible.

The first and second laws of thermodynamics may be employed to show that irreversible processes occurring during the operation of a heat engine reduce the fraction of thermal energy that ultimately is converted to work. Maximum energy conversion is obtained when entropy production by a heat engine is zero, i.e., when the engine operates reversibly. Such an ideal device is called a Carnot heat engine.

Extensive analyses of Carnot engines have been conducted to estimate the limiting performance of thermal power generation devices. A very important result relates the conversion efficiency, η, defined as

$$\eta = \frac{\Delta W}{\Delta Q_h}$$

to the absolute temperatures (in Kelvin or Rankine) of the thermal resource, T_h, and thermal sink, T_l. It can be shown that for a 2T Carnot heat engine operating cyclically,

$$\eta = 1 - \frac{T_l}{T_h}$$

Heat transfer between the Carnot engine and the thermal reservoirs must take place reversibly, i.e., over an infinitesimal temperature gradient. Clearly such heat transfer is impracticable in an actual device.

The high-temperature thermal source for OTEC is warm surface seawater. Even in the tropics, the temperature of this resource is not expected to exceed about 30°C or approximately 303 K. Since, from a practical standpoint, continuous cyclic operation of the OTEC heat engine is desirable, a cooler thermal sink must be identified. As is evident in the relationship for the Carnot efficiency, h of the cycle increases as the temperature of this sink is reduced. OTEC proposes to reject heat from the power generation system (engine) to cold seawater brought up from the depths of the ocean by a pipeline. Given current practical limitations on the required pumping system, the lowest temperature accessible probably lies around 4°C or 277 K. Hence, even under the best of operating conditions, an ideal heat engine would have an energy conversion efficiency of only

$$\eta \approx 1 - \frac{277}{303} = 0.086 \text{ or } 8.6\%$$

This means that over 90% of the thermal energy extracted from surface seawater is 'wasted'. This result can be compared with the Carnot efficiency of a typical state-of-the-art combustion steam power plant. Here, T_h is the supercritical boiler temperature that may be as high as 810 K, and the cooling water temperature, T_l, will be near ambient or 300 K. Hence,

$$\eta \approx 1 - \frac{300}{810} = 0.63 \text{ or } 63\%$$

While power cycles based on the combustion of fossil fuel enjoy significantly higher theoretical conversion efficiencies, the resource employed is limited, and undesirable by-products of the process, specifically combustion gas pollutant species, may damage the environment. OTEC, on the other hand, consumes, albeit relatively inefficiently, a resource that is constantly being renewed by the sun, and may be configured in a way that poses little threat to the environment.

It must be pointed out that the Carnot efficiency calculated above applies to an ideal, reversible cycle that exchanges heat with its surroundings over an infinitesimally small temperature gradient. An actual OTEC heat engine will transfer heat irreversibly and produce entropy at various points in the cycle. This will reduce further the fraction of energy extracted from the warm seawater that ultimately is converted to electrical power.

4.3 PERFORMANCE OF OTEC COMPONENTS

The majority of designs submitted for commercial OTEC systems propose to utilize modified versions of components found in conventional fossil fuel or geothermal electric power stations. These components include heat exchangers and turbomachinery such as compressors, turbines, and pumps. A review of means applied to assess the performance of these devices therefore is warranted. Performance will be quantified by means of global parameters such as isentropic or isothermal efficiencies and heat exchanger effectiveness. Although such an approach ignores the important physics of the operation of these devices, and restricts the level and extent of analysis possible, it is simple to apply and adequate for the present discussion.

Efficiencies and effectiveness are parameters that compare the ability of an actual device to perform a given task under specific operating conditions to the best theoretical performance possible by an ideal device (operating under the same conditions) that produces a minimum of entropy. Since methods to conduct thermodynamic analyses of ideal devices have been developed and are relatively simple to use, these global parameters afford a convenient means to assess the performance of power generation equipment. Values, charts, or equations for efficiencies and effectiveness are provided by manufacturers of devices to system integrators and analysts.

It has been suggested previously that the performance of an energy-related task will proceed most efficiently when all steps undertaken are reversible. This is reasonable if we accept irreversibilities, i.e., events or processes that produce entropy, degrade transfers or transformations of energy. Common sources of irreversibilities include friction (which manifests itself by the transformation of some organized macroscopic motion to random molecule motion, i.e., heating), mixing of different chemical species, heat transfer across a finite temperature difference, and spontaneous chemical reaction.

The three classes of turbomachinery that have been proposed for use in OTEC systems are pumps, turbines, and compressors. Pumps and compressors transform external power into an increase in fluid pressure. Pumps are used when the fluid is a liquid, compressors when the fluid is a gas. Turbines operate in reverse to pumps and compressors: they are designed to transform the internal energy of a working fluid, with a resultant pressure drop, into power that can be exported to the surroundings.

A major objective of pump and compressor design is to minimize power consumption for a given rise in fluid pressure. Hence, efficiencies compare the energy consumed by real and ideal devices operating between the same inlet fluid state and exit pressure, and handling the same mass flow rate. If the inlet state is denoted as 1 and the exit as 2 or 2s, where the s indicates that the desired pressure increase has occurred isentropically, then it can be shown that

$$\eta_{\text{pump}} = \eta_{\text{comp}} = \frac{W_s}{W_{\text{real}}} = \frac{h_{2s} - h_1}{h_2 - h_1}$$

where h_i is the enthalpy of the fluid at state i. These isentropic efficiencies will assume values that lie between 0 and 1. This expression has been derived assuming steady operation of the pump or compressor, negligible changes in kinetic and potential energy of the fluid, equilibrium at the inlet and outlet, and zero heat transfer between the fluid and the environment while passing through the device.

Although the theoretical basis of the relationship given above for h_{comp} is sound, it is found that the power consumed for a given compression ratio is actually minimized when the entropy production is zero and the temperature of the fluid remains constant. Such an ideal, isothermal compression process can only occur if heat transfer takes place reversibly from the fluid to the environment during the course of the compression. An isothermal compressor efficiency, similar in form to the isentropic efficiency given above, is often employed as an alternative performance parameter. In real systems, intercoolers are frequently used between compressor stages to approximate isothermal compression.

As expected, since maximizing power output for a given pressure drop is the primary objective of turbine design, the isentropic efficiency is defined as the inverse of the pump/compressor efficiency:

$$\eta_{turb} = \frac{W_{real}}{W_s} = \frac{h_2 - h_1}{h_{2s} - h_1}$$

where, again, 1 and 2 or 2s denote the states at the inlet and exit of the turbine, respectively. The assumptions underlying the above expression are identical to those used to analyze the pump and compressor. This isentropic turbine efficiency compares the performance of the actual and ideal turbine operating between the same inlet state and exit pressure. Note that since W_{real} is always $\leqslant W_s$, $0 \leqslant \eta_{turb} \leqslant 1$.

Heat transfer losses, fluid turbulence, and friction (viscous shear) in boundary layers reduce the isentropic efficiencies of turbomachinery components below the theoretical limit of 100%. Efficiencies of state-of-the-art, industrial turbines may attain value as high as about 90%; isentropic efficiencies of large, axial, and centrifugal compressors range between 70% and 85% while hydraulic pump efficiency rarely exceeds 65%.

Heat exchangers are devices employed to facilitate the exchange of energy, typically from one fluid to another, by heat transfer. Two major classes of these devices exist: (1) heat exchangers in which the fluids are separated by a solid boundary; (2) direct-contact devices in which the hot and cold fluid streams are allowed to mix.

Heat exchanger design focuses on maximizing the amount of thermal energy transferred between two fluid streams entering at a given pair of temperatures (or states). Practical objectives include minimizing the pressure drop, size, complexity, and cost of the heat exchanger.

Heat exchanger performance is quantified by means of a parameter known as the effectiveness, ϵ. Effectiveness compares the rate of heat transfer, Q_{real}, occurring between the hot and cold fluids within an actual device, and the maximum possible rate of heat transfer, Q_{max}, that would take place in a counterflow heat exchanger, with infinite heat transfer surface area, operating at the same mass flow rates and the same inlet states of the two fluids. In a counterflow heat exchanger, the two fluids flow in opposite directions, i.e., the colder fluid enters at the outlet of the warmer fluid flow path, and exits at its inlet.

The requirement that heat transfer can occur spontaneously only from a warmer body to a cooler body, imposed by the second law of thermodynamics, establishes the value of Q_{max}. If it is assumed that there are no heat losses to the environment, then it can be shown that Q_{max} is attained when the outlet temperature of the cooler fluid reaches the inlet temperature of the warmer fluid, for $m_c C_{pc} < m_h C_{ph}$, or, alternatively, when the outlet temperature of the warmer fluid reaches the inlet temperature of the cooler fluid, for $m_h C_{ph} < m_c C_{pc}$. Here, the subscripts c and h indicate conditions in the cooler and warmer fluid, respectively, and C_p is the constant pressure specific heat. It follows that

$$\epsilon = \frac{Q_{real}}{Q_{max}} \approx \frac{m_h C_{ph}(T_{h,in} - T_{h,out})}{\xi(T_{h,in} - T_{c,in})}$$

$$\epsilon = \frac{Q_{real}}{Q_{max}} \approx \frac{m_c C_{pc}(T_{c,in} - T_{c,out})}{\xi(T_{h,in} - T_{c,in})}$$

where the subscripts 'in' and 'out', respectively, denote conditions at the inlets and outlets and ξ is the smaller of $m_h C_{ph}$ and $m_c C_{pc}$. The approximate expressions for ϵ given above assume that changes in fluid enthalpy can be expressed as the product of a constant specific heat and a corresponding change in temperature.

The principal advantage of utilizing an effectiveness parameter in design calculations is that both the rate of heat transfer and the states of the two fluids exiting the heat exchanger may be determined explicitly, given a value of ϵ and the inlet states of the two fluid streams. Other methods, such as the LMTD (logarithmic mean temperature difference) analysis, require an implicit, iterative process to converge on the outlet states and Q_{real}.

4.4 FLUID DYNAMICS OF THE OTEC PIPELINE

Low energy conversion efficiencies of OTEC cycles require that thermal energy extracted from warm, surface seawater and rejected to cold, deep seawater exceed by nearly an order of magnitude the amount of work generated. As a consequence, huge flow rates of seawater will be required by commercial-scale OTEC plants.

It must be pointed out that the expression for the efficiency of the Carnot 2T heat engine given earlier does not consider the energy required to transport seawater to and from the power generating equipment. This pumping process is taken to be external to the 2T heat engine, and the energy consumed must be subtracted from the work output. As a consequence, the overall efficiency, both theoretical and real, of the OTEC cycle will be reduced further. Since the magnitude of this reduction scales directly with the amount of energy consumed in the pumping process, it is necessary for us to quantify this parasitic loss.

Rough estimates of the seawater pumping power may be obtained by employing the modified Bernoulli equation. Although relying on a high degree of empiricism, this expression, like those presented earlier for the component efficiencies, is simple to use and provides adequate results for many engineering analyses. If the subscripts 1 and 2 denote, respectively, the upstream and downstream ends of a section of circular pipe between which any number of valves, joints, or other passive flow operators may be installed, then, assuming zero heat transfer between the pipe and its surroundings,

$$(P_1 - P_2) + \frac{\rho(V_1^2 - V_2^2)}{2} + \rho g(z_1 - z_2) = \sum\left(\frac{fL\rho V^2}{2D}\right) + \sum\left(\frac{K\rho V^2}{2}\right)$$

In this equation, ρ is the liquid density, assumed constant, V is the average velocity at the relevant location, and z is the elevation. The two summations on the right-hand side account for friction losses occurring in the boundary layer adjacent to the pipe wall and fluid dynamic losses in valves, joints, etc. The friction factor, f, which depends on, among other things, pipe Reynolds number and wall roughness, is given in Moody diagrams. D is the inner diameter of the pipe, L is the length of pipe over which a set of values of f, V, and D are fixed, and K is an experimentally determined pressure drop factor.

The modified Bernoulli equation describes the pressure drop, $(P_1 - P_2)$, arising from a flow of liquid through a pipe. To sustain this flow, a pump may be required downstream of 2 to bring the liquid pressure up to ambient (or some other desired value). Given the liquid flow rate, P_2, and the desired pump exit pressure (i.e., the pressure ratio across the pump), an analysis utilizing an appropriate value of h_{pump} will determine the pumping power requirement. It should be observed that if the modified Bernoulli equation indicates that, for a given liquid flow rate and P_1, P_2 is equal to the local ambient at 2, then no external energy input is required.

If P_3 is the liquid pressure at the exit of a pump placed immediately downstream of location 2, and if fluid kinetic and potential energy changes across this pump can be neglected, then a good approximation of W_s/m, the power consumed per unit mass flow of liquid through an ideal pump, is

$$\frac{W_s}{m} \approx \frac{P_3 - P_2}{\rho}$$

Therefore,

$$\frac{W_{real}}{m} \approx \frac{P_3 - P_2}{\rho \eta_{pump}}$$

From the Bernoulli equation,

$$(P_3 - P_2) = (P_3 - P_1) + \frac{\rho(V_2^2 - V_1^2)}{2} + \rho g(z_2 - z_1)$$
$$+ \sum\left(\frac{fL\rho V^2}{2D}\right) + \sum\left(\frac{K\rho V^2}{2}\right)$$

Finally,

$$\frac{W_{real}}{m} \approx \left(\frac{1}{\eta_{pump}}\right)\left[\frac{P_3 - P_1}{\rho} + \frac{V_2^2 - V_1^2}{2} + g(z_2 - z_1)\right.$$

$$\left. + \sum\left(\frac{fLV^2}{2D}\right) + \sum\left(\frac{KV^2}{2}\right)\right]$$

A popular misconception about OTEC is that, due to the resisting gravitational force, an enormous amount of power must be consumed to bring seawater up to the surface from depths approaching 1000 m. The above equation for W_{real}/m shows that this is not true. Assuming that the density of seawater remains constant, pressure varies linearly with depth according to

$$P(z) = P_{atm} - \rho g z$$

where P_{atm} is the atmospheric pressure and z increases moving upward toward the ocean surface (where $z = 0$). If P_3 is taken to equal P_{atm}, then the terms $[(P_3 - P_1)/\rho]$ and $[g(z_2 - z_1)]$ cancel. Moreover, if the pipe diameters at 1 and 2 are the same, then $V_2 = V_1$. Under these circumstances, pumping power is required only to overcome frictional effects in the long, submerged pipeline.

4.5 OTEC SYSTEMS

Ocean thermal energy conversion (OTEC) extracts thermal energy from warm surface waters of tropical oceans to drive a 2T, cyclic heat engine. The heat engine transforms a portion of this energy to electrical power and rejects the balance to a colder, thermal sink. The thermal sink employed is seawater brought up from the ocean depths by means of a submerged pipeline. Although other techniques have been proposed to exploit the temperature gradient of tropical oceans to produce power, practical considerations suggest that near-term realization of OTEC lies in Rankine (closed-cycle) or Claude (open-cycle) plants.

Conventional heat engines employ a working fluid that circulates through different components, undergoing a series of transformations of state in the course of converting energy received from a thermal resource to work. Although there are many different ways to categorize heat engines, the primary means is by reference to the series of steps undertaken in the cycle. Variants of basic cycles often emerge as a result of efforts to improve efficiency or to accommodate new applications or energy resources.

Heat engines may operate in closed- or open-cycles. In a closed-cycle, the same working fluid is recirculated continuously through the system. Conventional combustion-driven or nuclear-driven steam powerplants are examples of closed-cycle heat engines. In an open-cycle device, the working fluid is not recirculated: a quantity of matter enters the engine, undergoes changes in state, and is exhausted to the environment. Including the environment with the heat engine as a single system closes this type of cycle; i.e., an open-cycle is closed by the environment. Examples of open-cycle heat engines include internal combustion engines and gas turbines used for stationary power production or propulsion.

In the past, small-scale, closed-cycle OTEC experimental plants have been constructed and operated successfully over short periods of time. These plants employ the same basic Rankine cycle used in most large electric power stations; however, ammonia or chlorofluorocarbons (CFCs), rather than steam, is used as the working fluid. Open-cycle OTEC employs steam produced by vaporization of warm seawater in a partial vacuum as the working fluid. This steam ultimately is condensed by heat transfer to cold seawater and discharged from the power generation system.

4.5.1 Closed-cycle OTEC

The basic, ideal Rankine cycle comprises four major steps:

1 isentropic compression (pumping) of a liquid;

2 constant pressure heat transfer from the thermal resource resulting in vaporization of the liquid;

3 isentropic expansion of the vapor through a turbine to generate power; and

4 constant pressure heat transfer to the thermal sink resulting in complete condensation of the working fluid.

The heat exchangers employed in steps (2) and (4) usually are called boilers and condensers, respectively.

A process representation for the ideal Rankine cycle, which depicts the changes in state of the working fluid as it progresses through the different steps, is shown in Figure 4.3. Following conventional engineering practice, the process has been plotted on the T–S (temperature–entropy) property plane of a hypothetical working fluid. The bell-shaped curve in the figure is known as the vapor dome. The vapor dome separates the liquid and vapor (gas) states of the fluid: points to the left of the dome are liquid, to the right vapor, and within its borders a mixture of saturated liquid and saturated gas (saturation implies boiling conditions). The segmented lines that horizontally cross the vapor dome are isobars, i.e., lines of constant pressure in the T–S plane.

A schematic diagram showing the major components used in a Rankine cycle is given in Figure 4.4. These four components include a pump, boiler, turbine, and condenser. State points marked on Figure 4.4 correspond to the numbers on the process representation.

64 OTEC THERMODYNAMICS

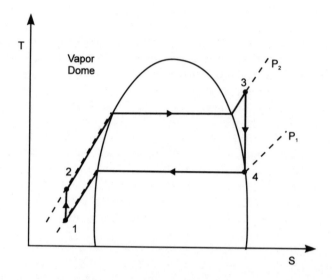

Figure 4.3 Ideal Rankine cycle process representation

In an actual Rankine heat engine, pressure drops occur in the boiler and condenser, and irreversibilities in the turbomachinery result in increases in the entropy of the working fluid passing through these devices. Referring to the definitions of component efficiencies, it can be shown that these effects increase the power consumed by the pumps and decrease the power generated by the turbine for each unit of energy received from the thermal resource. Net power output to the environment, i.e., the difference between

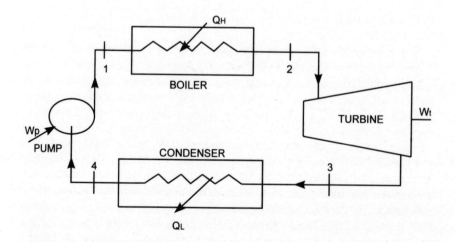

Figure 4.4 Rankine cycle components

power generated and power consumed by cycle parasitics, decreases along with cycle efficiency.

Rankine cycles are able, in theory, to produce non-zero net power, largely due to the fact that significantly less energy is required to increase the pressure of a liquid from some P_1 to P_2 than can be recovered as work when the same fluid expands as a vapor from the same P_2 back to P_1. Hence, the phase changes in the boiler and condenser are of critical importance to Rankine cycle operation.

In conventional steam power stations, the thermal resource is provided either by combusting a fossil fuel or through nuclear fission. Water from a river, ocean, or large artificial reservoir assumes the role of the thermal sink. Following the example of the Carnot 2T heat engine, the Rankine cycle is designed so that the working fluid, water, receives energy at as high an average temperature as practically possible, and rejects it at a temperature as close to that of the sink as the effectiveness of the condenser allows. Since saturation temperature varies directly with pressure, this requires that boiler pressures exceed atmospheric and that the steam side of the condenser operates at partial vacuum. High pressures and temperatures require heavy and expensive piping and components, while the in-leakage of air or coolant must be carefully guarded against at locations in the system where working fluid pressure drops below atmospheric. These practical considerations limit the efficiencies of steam power stations.

When the Rankine cycle is applied to the OTEC resource, where the difference between source and sink temperatures rarely exceeds 22°C (40°F), compared with up to 2000°C (3600°F) in combustion-driven systems, a different strategy must be applied. The major change is the substitution of a working fluid that boils and condenses at significantly lower temperatures than water at the same pressure. Factors that influence the selection of the working fluid include:

1 cost and availability,

2 compatibility with materials employed in piping, conventional turbo-machinery components, and heat exchangers,

3 toxicity, and

4 environmental hazard.

The leading candidates for the working fluid of closed-cycle OTEC plants are ammonia and several types of CFCs. Ammonia and CFCs are used extensively in refrigeration systems, which operate as Rankine cycles run in reverse, and also have been employed as working fluids in (relatively) low-temperature geothermal power plants. Their primary disadvantage is the environmental hazard posed by leakage of these substances. Ammonia is toxic in moderate concentrations and many types of CFCs attack and destroy the ozone layer of the atmosphere.

To understand the details of operation of closed-cycle OTEC, it is useful to consider a specific example. Figure 4.5 presents a flow diagram of a Rankine cycle plant employing ammonia as a working fluid. A heat and mass balance diagram for such a system designed to produce approximately 500 kW at the terminals of the electric generator driven by the turbine, and 270 kW net power, is shown in Figure 4.6. Table 4.1 summarizes the operation of this small, pre-commercial test plant that has been proposed for construction. Although a megawatt-size, commercial OTEC power system will enjoy economies of scale, the present example is not unrepresentative of anticipated performance.

As mentioned above, the closed-cycle OTEC system is identical to a simple Rankine steam power plant with the exception of the working fluid and the high-temperature thermal source. In the 500 kW closed-cycle plant depicted in Figure 4.6, warm surface seawater is pumped into the evaporator where heat transfer occurs to pressurize liquid ammonia. The pressure of the liquid ammonia is selected to ensure that boiling takes place several degrees below the temperature of the warm seawater. This accommodates the anticipated non-unity effectiveness of the evaporator. Ammonia vapor exiting the evaporator passes through a series of stop and control valves before entering the turbine. The vapor then expands through the turbine, turning a rotor connected to an electric generator. Ammonia vapor pressure at the outlet to the turbine corresponds to a saturation (condensation) temperature about 7°C higher than the temperature of the cold seawater. Cold seawater is brought up from a depth of 1000 meters and is pumped into the condenser where the ammonia vapor changes to liquid. This liquid ammonia is then pressurized by a pump and the cycle is repeated. Provisions to extract and purify the ammonia circulating in the power loop are included in the diagram along with a mixed discharge trench to collect and return the seawater effluent from the heat exchangers to the ocean.

It should be noted that the seawater pumps shown were sized assuming a land-based power plant utilizing bottom-mounted, submerged pipelines. A significant fraction of the power consumed by these pumps is expended to overcome friction in these pipelines. This loss scales with the length of the pipes and, hence, depends strongly on the bathymetry offshore of the plant. A gently sloping ocean floor will require a longer length of pipe to reach the required depth (i.e., access the desired sink temperature), incurring high power losses. The performance of land-based OTEC plants therefore varies with location. Floating plants, which mount power generating equipment on offshore barges, have been proposed to minimize this site-dependence. In floating plants, pipelines extend vertically downward.

In the present example, $T_h = 26.5°C$ and $T_l = 4.6°C$; a Carnot 2T heat engine therefore could attain a power conversion efficiency of 7.29%. Using typical values of efficiencies of commercially available equipment (e.g., $h_{turb} = 80\%$; h_{pump} (ammonia) $= 75\%$), thermodynamic analysis reveals that the efficiency of the real system, not including the seawater pumping losses, is

OTEC SYSTEMS 67

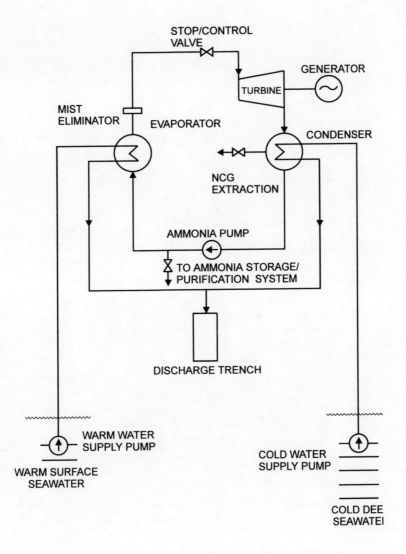

Figure 4.5 Closed-cycle OTEC flow diagram

68 OTEC THERMODYNAMICS

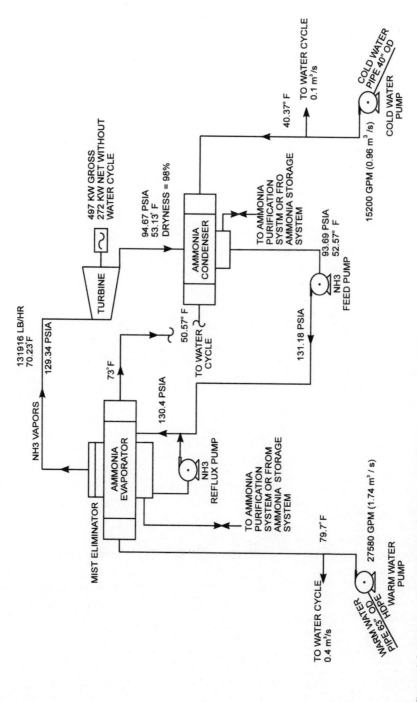

Figure 4.6 Heat and mass balance diagram of a 500 kW closed-cycle OTEC system

Table 4.1 Closed-cycle OTEC system summary

Warm water pipe outer diameter	1.6 meters (63 inches)
Cold water pipe outer diameter	1.0 meter (40 inches)
Mixed discharge pipe outer diameter	1.6 meters (63 inches)
Cold water pipe intake depth	1000 meters (3280 feet)
Gross power	497 kW
Net power	272 kW
Warm water flow	1.34 m^3/s (21 200 gpm)
Cold water flow	0.86 m^3/s (13 600 gpm)
Warm water supply temperature	26.5°C (79.7°F)
Cold water supply temperature	4.7°C (40.4°F)
Warm water return temperature	22.8°C (73.0°F)
Cold water return temperature	10.3°C (50.6°F)
Turbine inlet temperature	21.2°C (70.2°F)
Turbine inlet pressure	892 kPa (130 psia)
Turbine outlet temperature	11.7°C (53.1°F)
Turbine outlet pressure	653 kPa (95 psia)
Hydraulic loss in evaporator	2.5 meters (8.23 feet)
Hydraulic loss in condenser	2.6 meters (8.63 feet)
Cold water pipe friction loss	4.5 meters (14.75 feet)
Cold water pipe density head loss	0.5 meters (1.58 feet)
Warm water pipe friction loss	0.6 meters (1.93 feet)
Mixed discharge pipe friction loss	0.8 meters (2.68 feet)
Discharge trench water level	0.9 m above mean sea level
Warm water pump power	99.3 kW
Cold water pump power	114.6 kW
Ammonia pump power	10.7 kW

about 2.5%. Subtracting seawater pumping losses from the power generated by the turbine reduces the overall net efficiency to only 1.4%.

4.5.2 Open-cycle OTEC

The Claude or open-cycle proposes to use steam generated from the warm seawater as the working fluid. The steps of this cycle are:

1 flash evaporation of a fraction of the warm liquid by reduction of pressure below the saturation value corresponding to its temperature; this step may be perceived as heat transfer from the bulk of the liquid to the portion vaporized;

2 expansion of the vapor through a turbine to generate power;

3 heat transfer to the cold seawater thermal sink resulting in condensation of the working fluid; and

4 compression of the condensate and any residual non-condensible gases to pressures required to discharge them from the power generating system.

It is clear that the Claude cycle is very similar to the Rankine cycle with the exception that the state of the effluent at the end of step (4) need not match the state of the fraction of mass of the warm seawater that becomes the working fluid at the start of step (1).

Flash evaporation is a distinguishing feature of open-cycle OTEC. Flash evaporation involves complex heat and mass transfer processes. In the configuration most frequently proposed for open-cycle OTEC evaporators, warm seawater is pumped into a chamber through spouts designed to maximize heat and mass transfer surface area by producing a spray of the liquid. The pressure in the chamber is less than the saturation pressure of the warm seawater. Exposed to this low-pressure environment, water in the spray begins to boil. As in thermal desalination plants, the vapor produced is relatively pure steam. As steam is generated, it carries away with it its heat of vaporization. This energy comes from the liquid phase and results in a lowering of the liquid temperature and the cessation of boiling. Thus, as mentioned above, flash evaporation may be seen as a transfer of thermal energy from the bulk of the warm seawater to the small fraction of mass that is vaporized to become the working fluid. It is estimated that less than 0.5% of the mass of warm seawater entering the evaporator is converted into steam.

At 26.5°C, the approximate temperature of warm seawater, the saturation pressure, P_{sat}, of pure water is only 3464 Pa (0.50 psia). At a cold seawater temperature of, say, 4.7°C, saturation pressure is 852 Pa (0.12 psia). These two values of P_{sat} establish the approximate operating pressure range of an ideal, open-cycle OTEC system. Frictional pressure drops and non-ideal heat exchanger effectiveness will reduce this range in an actual plant. It is clear, however, that the evaporator, turbine, and condenser will operate in partial vacuum. This poses a number of practical concerns that must be addressed. First, the system must be carefully sealed to prevent in-leakage of atmospheric air that can severely degrade or shut down operation. Second, the specific volume of the low-pressure steam is very large compared to that of the pressurized working fluid used in closed-cycle OTEC. This means that components must have large flow areas to ensure that steam velocities do not attain excessively high values. Finally, gases such as nitrogen and carbon dioxide that are dissolved in seawater come out of solution in a vacuum. These gases will flow through the system with the steam but will not be condensed in the condenser. Since pressurization of a gas requires considerably more power than pressurization of a liquid, these non-condensable species increase the parasitic loss associated with the discharge step (4) of the process.

In spite of the aforementioned complications, the Claude cycle enjoys certain benefits from the selection of water as the working fluid. Water, unlike ammonia or CFCs, is non-toxic and environmentally benign. Moreover, since the evaporator produces desalinated steam, the condenser can be designed to yield a flow of potable, fresh water. In many potential OTEC sites in the tropics, potable water is a highly desired commodity that can be marketed to offset the price of OTEC-generated electricity.

A large turbine is required to accommodate the huge volumetric flow rates of low-pressure steam needed to generate any practical amount of electrical power. Although the last stages of turbines used in conventional steam power stations can be adapted to open-cycle OTEC operating conditions, it is widely accepted that existing technology limits the power that can be generated by a single open-cycle OTEC turbine module, comprising a pair of rotors, to about 4 MW. Unless significant effort is invested to develop new, specialized turbines (which may employ fiber-reinforced plastic blades in rotors having diameters in excess of 100 meters), increasing the gross power generating capacity of a Claude cycle plant above 4 MW will require multiple modules and incur an associated equipment cost penalty.

Condensation of the low-pressure working fluid leaving the turbine occurs by heat transfer to the cold seawater. This heat transfer may occur in a direct contact condenser (DCC), in which the seawater is sprayed directly over the vapor, or in a conventional shell-and-tube or plate-and-shell surface condenser that does not allow contact between the coolant and the condensate. DCCs have been proposed since they are relatively inexpensive and have good heat transfer characteristics due to the lack of a solid thermal boundary between the warm and cool fluids. Although surface condensers for OTEC applications may be expensive to fabricate and, possibly, difficult to maintain, they do permit the production of fresh water. Fresh water production is impossible with a DCC unless fresh water is substituted for seawater as the coolant in the DCC. In such an arrangement, the cold seawater sink is used to chill the fresh water coolant supply using a liquid/liquid heat exchanger.

Effluent from the low-pressure condenser must be returned to the environment. Liquid can be pressurized to ambient conditions at the point of discharge by means of a pump or, if the elevation of the condenser is suitably high, it can be compressed hydrostatically. Non-condensable gases, which include any residual water vapor, dissolved gases that have come out of solution, and air that may have leaked into the system, must be pressurized with a compressor. Although the primary role of the compressor is to discharge exhaust gases, it is usually perceived as the means to reduce pressure in the system below atmospheric. For a system that includes both the OTEC heat engine and its environment, the cycle is closed and parallels the Rankine cycle. The role of the Rankine cycle pump is assumed by the condensate discharge pump and the non-condensable gas compressor.

A heat and mass balance diagram for an open-cycle OTEC plant that produces about 1.1 MW of net electrical power and 1.26 million gallons/day of potable water is shown in Figure 4.7. This system was designed to accommodate the needs of a small community in a developing Pacific island nation. A separate stage that utilizes the seawater effluent from the power generation system evaporator and surface condenser has been included to

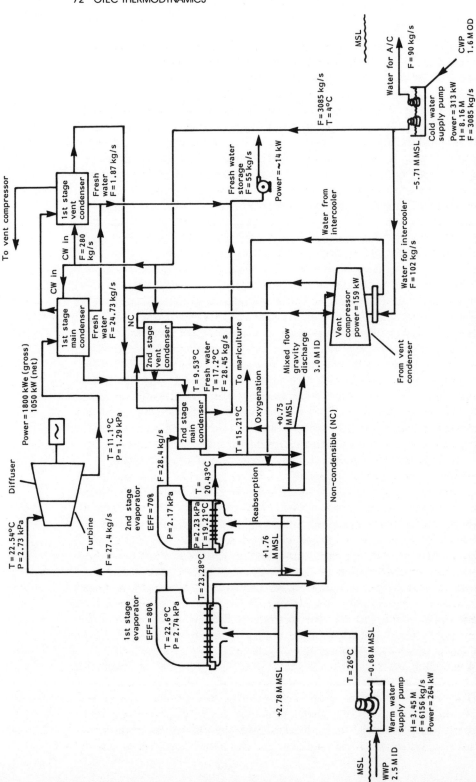

Figure 4.7 A heat and mass balance diagram for a commercial open-cycle OTEC plant that would produce about 1.1 MW of net electrical power

maximize fresh water production. This stage comprises a second flash evaporator and surface condenser only (no turbine). Such a fresh water production system could, in principle, be added to a closed-cycle plant to generate some fresh water.

Warm seawater at 26°C is discharged through spouts in a flash evaporation chamber maintained at approximately 2740 Pa. At this pressure, the seawater is about 3.4°C superheated. Calculations indicate that out of 6156 kg/s of seawater, only 27.4 kg/s of steam, 0.45% by mass, is produced. Transfer of the latent heat of vaporization to the steam results in a 2.7°C drop in the temperature of the seawater. The mixture of saturated steam and trace quantities of dissolved gases passes through a mist eliminator installed to remove tiny droplets of seawater entrained by the working fluid. The steam then passes through a single-stage, single-flow, axial turbine coupled to an electric generator. The turbine selected is a modified L-O (last row) stage as used in large nuclear power stations. The diameter of the rotor is 5.65 meters.

Although a diffuser is employed downstream of the rotor exit, care must be exercised not to expand to too low a pressure. Since in open-cycle OTEC, the working fluid will enter the turbine as either a saturated or a very slightly superheated vapor (due to a small pressure drop between the evaporator and turbine), some condensation can, and probably will, occur as this vapor expands to a lower pressure. Liquid droplets mixed with the steam will erode the rotor blades and degrade aerodynamic performance of the turbomachine.

After exiting the turbine diffuser the pressure of the working fluid is 1290 Pa, which corresponds to a condensation temperature of 10.7°C. Heat transfer to the 4°C cold seawater occurs in a surface condenser producing 26.6 kilograms of fresh water per second. Calculations indicate that about 3% of the water vapor will not be condensed. Finally, the non-condensable gases are removed from the system and are exhausted to the atmosphere by a vacuum compressor train that employs intercoolers to approximate the more efficient isothermal compression process. While a closed-cycle system will employ valves to throttle mass flow into the turbine and control its operation, it has been proposed that the compressor be used to perform this function in an open-cycle plant. Changing the rate at which non-condensables are removed by the compressor from the power generation loop will alter pressures in the system which, in turn, will affect steam production and applied torque (by the steam) to the turbine rotor. This strategy eliminates the fluid friction loss arising from control valves mounted in the steam path.

In Figure 4.7, the seawater pumps were sized for a land-based power plant sited at a location with similar bathymetry to that assumed for the 500 kW closed-cycle example. While open-cycle OTEC plants also may be mounted on offshore barges, means will have to be devised to transport both electricity and fresh water to shore. In the present example, the Carnot power conversion efficiency corresponding to $T_h = 26°C$ and $T_l = 4°C$ is 7.35%.

Thermodynamic analysis of the actual proposed cycle yields an efficiency of about 2.5%, if the seawater pumping losses are omitted, and about 1.6% if these losses are subtracted from the electrical power generated. Although there is a four-fold difference in the scale of the two example facilities and a second stage for water production has been included in the open-cycle system design, the performance results are quite similar.

The open- and closed-OTEC cycles are strategies to operate 2T heat engines that exploit the thermal gradient of tropical oceans to produce electric power. The small temperature difference between the proposed thermal source and sink limit maximum power conversion efficiency to about 7%. Irreversibilities in real devices and systems, and external parasitic losses associated with pumping the large volumes of seawater required to sustain operation, make operation at more than about 2% efficiency highly unlikely. Thermal performance of the two cycles is not expected to differ significantly for similar-size plants. In spite of its low efficiency, OTEC can be a viable alternative to conventional power generation methods since the energy resource it consumes is renewable and undesirable by-products are few.

The principal differences between open- and closed-cycle OTEC lie in the choice of working fluid. Closed-cycle systems propose to employ ammonia, CFCs, or some other high-pressure, low boiling point substance. Open-cycle OTEC uses low-pressure steam evaporated from the warm seawater. Because of their significantly higher operating pressures and, hence, lower volumetric flow rates, closed-cycle systems may employ compact components to produce the same amount of power as open-cycle systems. Assuming that means will be available efficiently to supply the required warm and cold seawater, closed-cycle plants can be designed to produce megawatts of net electrical power using existing turbomachinery and heat exchanger designs. On the other hand, practicable open-cycle OTEC systems with capacities in excess of one megawatt will probably require significant turbine development effort.

Although open-cycle OTEC plants suffer from large size and practical limitations on maximum power generation capacity, the working fluid employed poses no environmental hazard. Moreover, potable water produced in the condenser may be sold as a commodity to offset the cost of the OTEC-generated electrical power. The selection of either an open or closed OTEC cycle for a particular application will depend on the desired electrical capacity and the relative importance of power and potable water. Beyond a certain point, this decision will probably be driven by economic rather than technical factors.

SELF-ASSESSMENT QUESTIONS

1 Explain the operation of 1T and 2T heat engines. State and discuss the equations that describe the work produced by each type of engine.

2 Describe the two major classes of heat exchangers and state the practical objectives in heat exchanger design.

3 Describe Rankine and Claude cycles for heat engines and explain how these relate to open- and closed-cycle OTEC systems.

4 What are some of the complications of operating an evaporator, turbine, and condenser in the partial vacuum of an open-cycle OTEC system?

5 For the Rankine cycle shown in Figure 4.4, write an expression for the thermal efficiency in terms of properties at various points in the cycle.

6 Calculate the evaporator thermal effectiveness (ϵ) of the first stage evaporator for the open-cycle OTEC system shown in Figure 4.7 using the following equation:

$$\epsilon = \frac{T_{ww,in} - T_{ww,out}}{T_{ww,in} - T_{sat.\ steam}}$$

7 Discuss the differences between open- and closed-cycle OTEC. What are the advantages and disadvantages of each?

Answers

1 A heat engine receives energy from a thermal resource (or reservoir) by means of heat transfer. The second law of thermodynamics requires that heat transfer take place from a warmer body to a cooler body. If no energy is rejected from the heat engine to its environment by heat transfer, it is referred to as a 1T (for one thermal reservoir) heat engine. If heat is rejected to a cooler thermal sink, then it is called a 2T (for two thermal reservoirs) heat engine. For a 1T heat engine,

$$0 = DW - DQ_h + 0$$

and for a 2T heat engine

$$0 = (DW + DQ_1) - DQ_h + 0$$

DW and DQ are the total energy transfers, which take place over the course of a cycle, between the engine and its surroundings, by work and heat, respectively. Rearranging terms we have expressions for the work produced:

$$DW = DQ_h \quad \text{(1T heat engine)}$$

or

$$DW = DQ_h - DQ_1 \quad \text{(2T heat engine)}$$

For an 1T heat engine operating cyclically, all energy received from the thermal resource is converted to work. For a 2T heat engine, only a portion of this energy is converted to work, the balance being reject to the thermal sink.

2 Two major classes of heat exchangers exist: (1) devices in which the fluids are separated by a solid boundary; and (2) direct-contact devices in which the hot and cold fluid streams are allowed to mix. Heat exchanger design focuses on maximizing the amount of thermal energy transferred between two fluid streams entering at a given pair of temperatures (or states). Practical objectives include minimizing the pressure drop, size, complexity, and cost of the heat exchanger.

3 The basic Rankine cycle comprises four major steps: (1) isentropic compression (pumping) of a liquid; (2) constant pressure heat transfer from the thermal resource resulting in vaporization of the liquid; (3) isentropic expansion of the vapor through a turbine to generate power; and (4) constant pressure heat transfer to the thermal sink resulting in complete condensation of the working fluid. The Rankine cycle is employed in closed-cycle OTEC.

The Claude, or open-cycle, proposes to use steam generated from warm seawater as the working fluid. The steps of this cycle are: (1) flash evaporation of a fraction of the warm liquid by reduction of pressure below the saturation value corresponding to its temperature; this step may be perceived as heat transfer from the bulk of the liquid to the portion vaporized; (2) expansion of the vapor through a turbine to generate power; (3) heat transfer to the cold seawater thermal sink resulting in condensation of the working fluid; and (4) compression of the condensate and any residual non-condensable gases to pressures required to discharge them from the power generating system.

4 Complications of operation in a partial vacuum:

(a) The system must be carefully sealed to prevent in-leakage of atmospheric air that can degrade or shut down operation.

(b) The specific volume of the low-pressure steam is very large compared to that of the pressurized working fluid used in closed-cycle OTEC.

(c) Gases that are dissolved in seawater come out of solution in a vacuum. These gases will flow through the system with the steam but will not be condensed in the condenser and will increase the parasitic loss associated with the discharge step of the process.

5 Cycle thermal efficiency $= \dfrac{(h_2 - h_1) - (h_3 - h_4)}{h_2 - h_1}$

6 Evaporator effectiveness $= \dfrac{26 - 23.28}{26 - 22.6} = 80\%$

7 The principal difference between open- and closed-cycle OTEC is the choice of the working fluid. The closed-cycle working fluid is ammonia or CFCs (low boiling point substance); advantages include:

(a) employs compact components to produce the same amount of power as open-cycle OTEC systems;

(b) closed-cycle plants can be designed to produce megawatts of net electrical power using existing turbomachinery and heat exchanger designs.

The open-cycle working fluid is low pressure steam evaporated from warm seawater; its advantages include:

(a) working fluid imposes no environmental hazard;

(b) fresh water produced in condenser may be sold as a commodity.

References

COMMITTEE ON ALTERNATE ENERGY SOURCES FOR HAWAII OF THE STATE ADVISORY TASK FORCE ON ENERGY POLICY, *Alternate Energy Sources for Hawaii*, Hawaii Natural Energy Institute, University of Hawaii, Honolulu, 1975.

Electric Power Research Institute, AP-4921, *Ocean Energy Technologies: The State of the Art*, Prepared by Massachusetts Institute of Technology, Department of Ocean Engineering, Project 1348–28, Final Report, November 1986.

HAGERMAN, G. and HELLER, T., Wave energy: a survey of twelve near-term technologies, *Proceedings of the International Renewable Energy Conference*, Honolulu, Hawaii, September 1988.

HAY, G. A., Ocean wave energy technology status, resources and applications for northern California, *Globe '90*, Vancouver, British Columbia, March 1990.

LISSAMAN, P. B. S., The Coriolis program, *Oceanus*, Vol. 22, No. 4, Winter 1979/80, pp. 23–28.

MONNEY, N. T., Ocean energy from salinity gradients, *Ocean Energy Resources*, Energy Technical Conference, American Society of Mechanical Engineers, Houston, Texas, September 1977, pp. 33–42.

NATIONAL OCEANIC AND ATMOSPHERIC ADMINISTRATION OFFICE OF OCEAN AND COASTAL RESOURCE MANAGEMENT, *Ocean Thermal Energy Conservation*, Report to Congress: Fiscal Year 1983, Washington, DC, 1984.

PENNY, T. R. and BHARATHAN, D., Power from the sea, *Scientific American*, vol. 256, No. 1, January 1987, pp. 86–92.

REYNOLDS, W. C. and PERKINS, H. C., *Engineering Thermodynamics*, 2nd edn., McGraw-Hill, New York, 1977.

ROGERS, L. and TRENKA, A., An update on the U. S. DOE energy program, *Proceedings of the Ocean Energy Recovery 1st International Conference*, WW Division of ASCE, Honolulu, Hawaii, 1989.

SARIS, E. C., SCHOLTEN, W. B. KERNER, D. A. and LEWIS, L. F., Overview of international ocean energy activities, in *Ocean Energy Recovery*, edited by H. J. Krock, ICOER Proceedings, Honolulu, Hawaii, November 1989, pp. 1–34.

SOLAR ENERGY RESEARCH INSTITUTE (SERI), *Ocean Thermal Energy Conversion – an Overview*, SERI/SP–220–3024, November 1989.

TAKAHASHI, P. and TRENKA, A., Ocean thermal energy conversion: its promise as a total resource system, *Energy*, Vol. 17, No. 7, 1992, pp. 657–668.

TAKAHASHI, P., KINOSHITA, C., ONEY, S. and VADUS, J., Facilitating technology for fuel production and energy-enhanced products, Chapter 12 in *Ocean Energy Recovery: The State of the Art*, edited by Richard J. Seymour, ASCE, New York, 1992.

TAYLOR, R. H., *Alternative Energy Sources*, Adam Hilger, Bristol, 1983, pp. 141–144.

TWIDELL, J. W. and WEIR, A. D., *Renewable Energy Resources*, E. & F. N. Spon, London, 1986, p. 349.

VADUS, J. R., Global environment and engineering, *10th Ocean Engineering Symposium*, Nihon University, Tokyo, Japan, January 1991.

VEGA, L. A. and TRENKA, A. R., *Near market potential for OTEC in the Pacific Basin*, Pacific International Center for High Technology Research, Honolulu, Hawaii, 1989.

WARNOCK, J. G., Energy from the oceans–a review of recent and proposed tidal power installations, *World Federation of Engineering Organizations Technical Congress*, Vancouver, British Columbia, May 1987.

Index

Africa, 4
Agriculture, 2, 6, 8, 9, 12
Air-conditioning, 2, 6, 8, 9, 12, 41
American Samoa, 7, 31
Ammonia, 4, 35, 42, 63, 65, 66, 70, 74
Anderson, J. Hilbert, 4
Apollo Project, 9
Aquaculture, 8
Artificial upwelling, 2, 9
Australia
   current energy, 21

Belau, 7
Biomass
   marine, 2, 9
Brazil, 4
By-products
   see *Agriculture, Air-conditioning,*
      Desalinated water, Mariculture,
      Multiple-product OTEC,
      Refrigeration

Canada
   current energy, 23
   tidal energy, 16
Carnegie-Mellon University, 4
Carnot, 56, 57, 60, 65, 66
China
   tidal energy, 16
Chlorofluorocarbons, 63, 65, 70, 74
Claude, Georges, 4
Claude cycle OTEC, 62, 69, 70, 71
Closed-cycle OTEC

construction of, 6
demonstration plant, 38
description of, 4, 66
heat exchangers, 4
and hybrid-cycle, 6
Cold water
   depth, 2, 66
   nutrients, 2, 4, 6, 9, 11, 41
   pathogens, 2, 4, 6
   temperature of, 2
   thermodynamics, 60
Cold water pipes
   demonstration project, 38
   flexible membrane, 11
   at Kochi, 8
   large-diameter, 11
   as major cost factor, 7
   at Natural Energy Laboratory, 6, 8
   pumps, 11
Condenser
   in hybrid-cycle, 6
Cook Islands, 7
Cuba, 4
Current energy
   Coriolis Project, 45
      construction cost, projected, 45
      electricity generation, projected, 45
      environmental impact, 45
   Darrieus-type vertical axis turbine, 23
   extractable power, projected, 45
   generation of, 22
   in Gulf Stream, 23
   maximum power formula, 23

Current energy (*Contd.*)
  progress of, 2
  turbine, 23

d'Arsonval, Jacques Arsene, 2
Desalinated water, 2, 9, 12
  cost of, 8, 41
  from demonstration plant, 38
  from flash desalination plant, 8
  from hybrid-cycle OTEC, 6, 7, 34
  as a multiple-product, 4, 6
  from open-cycle OTEC, 4, 7, 34, 41, 70, 71, 73
  from reverse osmosis plants, 8
  from wave energy, 22, 45
Direct contact condenser, 71

Electric Power Research Institute, 38
Electricity from OTEC, 2, 4, 6, 9, 34, 41
  cost-competitiveness for Pacific islands, 7
  cost per kilowatt hour, 7, 38
  levelized cost, 38
Environmental impact, 2, 35, 41, 42, 65, 70
Evaporator
  flash, in open-cycle, 4, 69, 70
  flash, in hybrid-cycle, 6, 35
Exclusive Economic Zone, 35

Financing of OTEC, 11, 35
  estimated cost, 38
Fisheries, 2, 9
Floating platforms or plantships, 2, 9, 11, 35, 38, 66, 73
France
  La Rance power station, 18
  OTEC design, 4
  tidal energy, 2, 18
Freon, 4, 42
Fresh water
  see *Desalinated water*

Guam, 7, 35

Hawaii
  OTEC experiment site, 4, 6, 8, 9, 41
  State of, 6, 31
Hawaii Ocean Science and Technology Park, 4, 6, 8

Heat exchanger
  for air-conditioning, 8
  in closed-cycle OTEC, 4
  in open-cycle OTEC, 4
  thermodynamics of, 57, 59, 60, 61
Heat transfer, 50, 51, 52, 54, 56, 59, 60, 69, 71, 73
Heronemus, William E., 4
Hurricanes, 11
Hybrid-cycle OTEC, 6, 34, 35

India
  wave energy, 20
Integrated OTEC, 9
International Energy Agency, 20
Islands and OTEC, 2, 6, 7, 11, 33, 34, 35, 73

Japan
  Government of, 6
  Kochi Artificial Upwelling Laboratory, 8
  multiple-product OTEC, 6
  wave energy research, 20

Kiribati, Republic of, 7

Mariculture, 2, 6, 8, 9, 11, 12, 41
Mexico
  current energy, 25
Mini-OTEC, 4, 41
Mining, 9
Multiple-product OTEC, 4, 6, 14

National Aeronautics and Space Administration, 11
National Oceanic and Atmospheric Administration, 11
National Science Foundation, 4
Natural Energy Laboratory, 4, 6, 8, 41
Northern Mariana Islands, Commonwealth of, 7, 35
Norway
  wave energy, 20

Oceans
  as alternate energy source, 1
Ocean energy
  total power available, 2
Oil embargo, 4

Open-cycle OTEC
  demonstration plant, 38
  description, 4, 69, 70
  and hybrid-cycle, 6
  markets for, 30
  net power-producing experiment (NPPE), 6, 12
  thermodynamics and, 69, 70
Open ocean
  biomass plantations, 2
  fisheries, 2
OTEC
  demonstration plant, 38
  impediments to commercialization, 11, 12, 38, 74
  progress of, 2
  markets for, 33
  market penetration scenarios, 34, 35
  shore-, land-based plants, 7, 35, 38, 41
OTEC-1, 4

Pacific International Center for High Technology Research (PICHTR), 6, 7, 8, 11
Pacific and Asia region
  OTEC survey, 6
Philippines
  current energy, 23
Pohnpei, 7

Rankine cycle OTEC, 56, 62, 63, 64, 65, 66, 70, 71
Refrigeration, 2, 6, 8, 12, 65

Salinity gradient energy
  diagram of power unit, 28, 32
  electricity cost, projected, 46
  generation of, 23, 25
  membrane, 25, 46
  progress of, 2
Seawater
  applications in Hawaii, 4
  temperature difference, 2, 35, 56, 62
Solar energy, 2, 54
Soviet Union
  tidal energy, 18
  wave energy, 22
Sweden
  wave energy, 22

Tahiti, 31
Taiwan, 35
Thermodynamics, in OTEC
  boilers, 63, 65
  compressors, 57, 58, 63, 73
  convection, 52, 54
  definition, 50
  enthalpy, 51
  entropy, 50, 53, 54, 56, 58, 63, 64
  equilibrium, 50
  first law of, 52, 53, 56
  heat engine, 55, 56, 57, 60, 62, 65, 74
  laws of, 50
  macroscopic components, 50, 51
  microscopic components, 50, 51
  and pumps, 57, 58, 60, 61, 63
  in salinity gradient energy, 25
  second law of, 53, 54, 56
  work, 50, 52, 56
Tidal energy
  construction cost, projected, 42
  electricity cost, projected, 42
  electricity generation, 17, 18
  environmental impact, 42, 44
  geographic considerations, 18
  progress of, 2
  sea level, 19
Tonga, 7
Turbine
  in current energy systems, 23, 45, 46
  Darrieus-type vertical axis, 23
  design of, 11
  generator, 4
  low-pressure, 4, 71
  in salinity gradient systems, 25
  thermodynamics, 57, 58, 63, 66, 69
  in tidal energy systems, 18
  in wave energy systems, 20

United Kingdom
  tidal energy, 42
  wave energy, 20, 22
United States
  current energy, 23
US Department of Energy, 6
US Department of State
  OTEC study, 35
University of Massachusetts, 4

Vacuum, 4

Wave energy
   construction cost, projected, 44
   electricity cost, projected, 44
   environmental impact, 44, 45
   generation of, 19
   heaving buoy, 22
   oscillating water column, 20
   progress of, 2
   tapered channel, 20
   technologies, 20
   wave theory, 24
Western Samoa, 7

Zener, Clarence, 4